MOLECULAR AND PARTICLE MODELLING OF LAMINAR AND TURBULENT FLOWS

MOLECULAR AND PARTICLE MODELLING OF LAMINAR AND TURBULENT FLOWS

by

Donald Greenspan

University of Texas at Arlington, USA

 World Scientific

NEW JERSEY · LONDON · SINGAPORE · BEIJING · SHANGHAI · HONG KONG · TAIPEI · CHENNAI

Published by

World Scientific Publishing Co. Pte. Ltd.

5 Toh Tuck Link, Singapore 596224

USA office: 27 Warren Street, Suite 401-402, Hackensack, NJ 07601

UK office: 57 Shelton Street, Covent Garden, London WC2H 9HE

Library of Congress Cataloging-in-Publication Data
Greenspan, Donald.
  Molecular and particle modelling of laminar and turbulent flows / by Donald Greenspan.
    p. cm.
  Includes bibliographical references and index.
  ISBN-13 978-981-256-096-4 (alk. paper)
  ISBN-10 981-256-096-3 (alk. paper)
    1. Turbulence--Mathematical models. 2. Laminar flow--Mathematical models. 3. Water
  vapor transport--Mathematical models. 4. Vapors--Mathematical models. I. Title.

QA913.G68 2005
532'.052'015118--dc22

2004066158

British Library Cataloguing-in-Publication Data
A catalogue record for this book is available from the British Library.

Typeset by Stallion Press
Email: enquiries@stallionpress.com

Printed in Singapore

Preface

Turbulence is the most fundamental and, simultaneously, the most complex form of fluid flow. By necessity, because of the myriad phenomena exhibiting turbulent behavior, this monograph focuses primarily, but not always, on single type problem, cavity flow. However, because an understanding of turbulence requires an understanding of laminar flow, both will be explored.

Groundwork is laid by careful delineation of the necessary physical, mathematical, and numerical requirements for the studies which follow, and include discussions of N-body problems, classical molecular mechanics, dynamical equations, and the leap frog formulas for very large systems of second order ordinary differential equations. Molecular systems are then studied in both two and three dimensions, while particle systems, that is, systems which use lump massing of molecules, are studied in only two dimensions.

All calculations are limited to a personal scientific computer, in our case a Digital Alpha 533, so that the methods can be utilized readily by others. Our choice of the Alpha 533 is motivated out of the desire to maximize accuracy and minimize computer time. This computer has a 64 bit word built into the hardware. Three dimensional calculations, which are restricted to Chapter 5 only, required several "tricks" in order to enable their completion in a reasonable time, and these will be described in Chapter 5.

Though molecular simulations are of interest in themselves, they are also completely consistent with the current surge of interest in nano physics and with our belief that the mechanisms of turbulence are on the molecular level. Nevertheless, extension into the large is also of great interest, and it is for this purpose that we develop particle mechanics.

Though Sec. 2.4 is essential reading for all the compuations described in Chapters 2–7, these chapters are, in general, relatively independent.

Finally, it should be observed that very often velocity fields for various figures, throughout, had to be rescaled for graphical clarity.

Contents

Chapter 1

Mathematical, Physical, and Computational Preliminaries

1.1. The N-Body Problem

The fundamental mathematical problem considered throughout this book is a non continuum problem called the N-body problem, which is described in complete generality as follows. In cgs units and for $i = 1, 2, \ldots, N$, let P_i of mass m_i be at $\vec{r}_i = (x_i, y_i, z_i)$, have velocity $\vec{v}_i = (v_{i,x}, v_{i,y}, v_{i,z})$, and have acceleration $\vec{a}_i = (a_{i,x}, a_{i,y}, a_{i,z})$ at time $t \geq 0$. Let the positive distance between P_i and $P_j, i \neq j$, be $r_{ij} = r_{ji} \neq 0$. Let the force on P_i due to P_j be $\vec{F}_{ij} = \vec{F}_{ij}(r_{ij})$, so that the force depends only on the *distance* between P_i and P_j. Also, assume that the force $\vec{F}_{ji}$ on P_j due to P_i satisfies $\vec{F}_{ji} = -\vec{F}_{ij}$. Then, given the initial positions and velocities of all the $P_i, i = 1, 2, 3, \ldots, N$, the general N-body problem is to determine the motion of the system if each P_i interacts with all the other P_j's in the system.

The prototype N-body problem was formulated in about 1900. In it the P_i were the Sun and the then known eight planets, and the force on each P_i was gravitational attraction. This problem is exceptionally difficult for $N \geq 3$. The additional difficulties with the problems to be considered in this book arise from the fact that we will be concerned with the interactions of molecules and particles (molecular aggregates), for which the forces are more complex than gravitation.

1.2. Classical Molecular Potentials

Classical molecular forces behave, in general, as follows (Feynman, Leighton and Sands (1973)). When two *close* (to be made precise shortly) molecules are pulled apart, they attract. When pushed together, they repel. And

1

the force of repulsion is of a greater order of magnitude than the force of attraction.

Perhaps the most important exception to the behavior just described is the basic fluid of all living matter, namely, liquid water, and this will be discussed later.

Example. Consider two hypothetical molecules P_1, P_2 on an X axis as shown in Fig. 1.1. Let P_1 be at the origin and let P_2 be R units from $P_1, R > 0$. Let $\vec{F}$ be the force P_1 exerts on P_2 and let F be the magnitude of $\vec{F}$. Suppose

$$F = \frac{1}{R^{13}} - \frac{1}{R^7}.$$

Then, if $R = 1, F = 0$, and the molecules are in equilibrium. If $R > 1$, say, $R = 2$, then

$$F = \frac{1}{2^{13}} - \frac{1}{2^7} < 0,$$

so that $\vec{F}$ acts toward the origin, which corresponds to attraction. If $R < 1$, say, $R = 0.1$, then

$$F = \frac{1}{0.1^{13}} - \frac{1}{0.1^7} > 0,$$

so that $\vec{F}$ acts away from the origin, which corresponds to repulsion. Note that F is unbounded as R converges to zero, so that as R converges to zero the interaction of the two molecules can be extremely volatile.

There are a variety of classical molecular potentials for the interactions of molecules and from these classical molecular force formulas can be derived (Hirschfelder, Curtiss and Bird (1967)). There are, for example, Buckingham, Lennard–Jones, Morse, Slater–Kirkwood, Stockmayer, Sutherland, and Yntema–Schneider potentials. The potential which has received the most attention is the Lennard–Jones potential, that is,

$$\phi(r_{ij}) = 4\epsilon \left[\frac{\sigma^{12}}{r_{ij}^{12}} - \frac{\sigma^6}{r_{ij}^6} \right] \text{ erg,}$$

some examples of which can be found in Table 1.1.

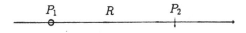

Fig. 1.1.   Force interaction.

Let us turn attention first to a Lennard–Jones potential for argon vapor. We examine this first because it is thought that for argon, the Lennard–Jones potential is *quantitatively* accurate (Koplik and Banavar (1998)). The potential is

$$\phi(r_{ij}) = (6.848)10^{-14} \left[\frac{3.418^{12}}{r_{ij}^{12}} - \frac{3.418^6}{r_{ij}^6} \right] \text{erg} \quad \left(\frac{\text{grcm}^2}{\text{sec}^2} \right) \qquad (1.1)$$

in which r_{ij} is measured in angstroms (Å). The force $\vec{F}_{ij}$ exerted on P_i by P_j is then

$$\vec{F}_{ij} = (6.848)10^{-14} \left[\frac{12(3.418)^{12}}{r_{ij}^{13}} - \frac{6(3.418)^6}{r_{ij}^7} \right] (10)^8 \frac{\vec{r}_{ji}}{r_{ij}} \text{dynes} \quad \left(\frac{\text{grcm}}{\text{sec}^2} \right).$$

$$(1.2)$$

Note that in deriving (1.2) from (1.1), one must use the chain rule

$$F_{ij} = -\frac{d\phi(r_{ij})}{dR} = -\frac{d\phi(r_{ij})}{dr_{ij}} \frac{dr_{ij}}{dR}$$

and the fact that R cm $= 10^8 R$ Å, that is, $r_{ij} = 10^8 R$. Hence, (1.2) reduces readily to

$$\vec{F}_{ij} = \left[\frac{209.0}{r_{ij}^{13}} - \frac{0.06551}{r_{ij}^7} \right] \frac{\vec{r}_{ji}}{r_{ij}} \text{dynes} \quad \left(\frac{\text{grcm}}{\text{sec}^2} \right). \qquad (1.3)$$

Table 1.1. Lennard–Jones (6–12) Potentials
$\phi(r_{ij}) = 4\epsilon \left[\frac{\sigma^{12}}{r_{ij}^{12}} - \frac{\sigma^6}{r_{ij}^6} \right]$ erg, $k = (1.381)10^{-16}$.

Gas	ϵ/k (°K)	σ (Å)
Ar	124	3.418
Ne	35.7	2.789
CO	110	3.590
CO_2	190	3.996
NO	119	3.470
CH_4	137	3.822
SO_2	252	4.290
F_2	112	3.653
Cl_2	357	4.115
C_6H_6	440	5.270
Air	97.0	3.617

Note also that

$$F_{ij} = \|\vec{F}_{ij}\| = \left[\frac{209.0}{r_{ij}^{13}} - \frac{0.06551}{r_{ij}^{7}} \right],$$

so that $F_{ij}(r_{ij}) = 0$ implies that $r_{ij} = 3.837\,\text{Å}$, which is the *equilibrium* distance.

1.3. Molecular Mechanics

Molecular mechanics is the simulation of molecular interaction as an N-body problem using classical molecular potentials and Newtonian mechanics. Most fluid studies using molecular mechanics are concerned with the *physics* of fluids and are concerned with flows at low Reynolds numbers. Recall then that, at *low* Reynolds number, the temperature T on the Kelvin scale of a molecule of mass m gr and speed v cm/sec is given in two dimensions by

$$kT = \frac{1}{2}mv^2 \tag{1.4}$$

and in three dimensions by

$$\frac{3}{2}kT = \frac{1}{2}mv^2, \tag{1.5}$$

in which k is the Boltzmann constant $(1.381)10^{-16}\,\text{erg deg}^{-1}$. Also recall that T degrees Kelvin and C degrees centigrade are related by

$$T = C + 273. \tag{1.6}$$

But, perhaps most importantly, it must be understood clearly at the outset that statistical mechanics concepts and formulas, including temperature, do *not* apply to turbulent flows, which will be our primary concern and which occur at *high* Reynolds numbers (Batchelor (1960), Bernard (1998), Koplik and Banavar (1998), Speziale and So (1998)). Related details and considerations will be discussed shortly.

1.4. The Leap Frog Formulas

Classical molecular force formulas require small time steps in any numerical simulation in order to yield physically correct results for the effect of repulsion, which is unbounded when the distance between the molecules is close to zero. Because we are restricted physically to small time steps and because the number of equations is usually exceptionally large, Runge–Kutta and Taylor expansion methods, for example, prove to be unwieldy for related problems. Hence, in this section we will develop a simplistic, efficient, but low order method, called the Leap Frog Method, for molecular mechanics simulations and it is described as follows.

Choose a positive time step h and let $t_k = kh, k = 0, 1, 2, \ldots$ For $i = 1, 2, 3, \ldots, N$, let P_i have mass m_i and at t_k let it be at $\vec{r}_{i,k} = (x_{i,k}, y_{i,k}, z_{i,k})$, have velocity $\vec{v}_{i,k} = (v_{i,k,x}, v_{i,k,y}, v_{i,k,z})$, and have acceleration $\vec{a}_{i,k} = (a_{i,k,x}, a_{i,k,y}, a_{i,k,z})$. The leap frog formulas, which relate position, velocity and acceleration are

$$\frac{\vec{v}_{i,\frac{1}{2}} - \vec{v}_{i,0}}{\frac{1}{2}h} = \vec{a}_{i,0}, \qquad \text{(Starter)} \tag{1.10}$$

$$\frac{\vec{v}_{i,k+\frac{1}{2}} - \vec{v}_{i,k-\frac{1}{2}}}{h} = \vec{a}_{i,k}, \qquad k = 1, 2, 3, \ldots \tag{1.11}$$

$$\frac{\vec{r}_{i,k+1} - \vec{r}_{i,k}}{h} = \vec{v}_{i,k+\frac{1}{2}}, \qquad k = 0, 1, 2, 3, \ldots, \tag{1.12}$$

or, explicitly,

$$\vec{v}_{i,\frac{1}{2}} = \vec{v}_{i,0} + \frac{1}{2}h\,\vec{a}_{i,0}, \qquad \text{(Starter)} \tag{1.13}$$

$$\vec{v}_{i,k+\frac{1}{2}} = \vec{v}_{i,k-\frac{1}{2}} + h\,\vec{a}_{i,k}, \qquad k = 1, 2, 3, \ldots \tag{1.14}$$

$$\vec{r}_{i,k+1} = \vec{r}_{i,k} + h\,\vec{v}_{i,k+\frac{1}{2}}, \qquad k = 0, 1, 2, 3 \ldots. \tag{1.15}$$

Note that (1.11) and (1.12) are two point central difference formulas. The name *leap frog* derives from the way position and velocity are defined at alternate, sequential time values. As shown in Fig. 1.2, the r values are defined at the times $t_0, t_1, t_2, t_3, \ldots$, while the v values are defined at the times $t_{\frac{1}{2}}, t_{\frac{3}{2}}, t_{\frac{5}{2}}, t_{\frac{7}{2}}, \ldots$, and indeed the figure is symbolic of the children's game *leap frog*.

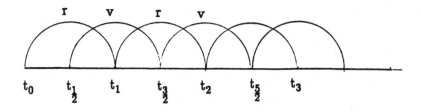

Fig. 1.2. Leap frog.

In formulas (1.13)–(1.15), the values $\vec{a}_{i,k}$ are determined from

$$\vec{a}_{i,k} = \frac{\vec{F}_{i,k}}{m_i}.$$

Example. For illustrative purposes only, let us consider the following simple example. On an X axis, let P_1, P_2, with respective masses $m_1 = 2$, $m_2 = 1$ be located initially at $x_{1,0} = 0, x_{2,0} = 1$ and have initial velocities $v_{1,0} = -1, v_{2,0} = 3$. Let $h = 0.1$ and assume that the forces on P_1 and P_2 at t_k are given by

$$F_{1,k} = \left(-\frac{1}{|x_{2,k}-x_{1,k}|^3} + \frac{2}{|x_{2,k} - x_{1,k}|^6} \right) \frac{x_{1,k} - x_{2,k}}{|x_{2,k} - x_{1,k}|} \qquad (1.16)$$

$$F_{2,k} = -F_{1,k}. \qquad (1.17)$$

Then (1.13)–(1.15) yield for P_1

$$v_{1,\frac{1}{2}} = v_{1,0} + (0.05)\,a_{1,0}, \qquad \text{(Starter)}$$

$$v_{1,k+\frac{1}{2}} = v_{1,k-\frac{1}{2}} + (0.1)\,a_{1,k}, \quad k = 1,2,3,\ldots$$

$$x_{1,k+1} = x_{1,k} + (0.1)\,v_{1,k+\frac{1}{2}}, \quad k = 0,1,2,3\ldots,$$

or, equivalently,

$$v_{1,\frac{1}{2}} = -1 + (0.05)\left(\frac{1}{2}\right) F_{1,0} = -1 + (0.025) F_{1,0} \qquad (1.18)$$

$$v_{1,k+\frac{1}{2}} = v_{1,k-\frac{1}{2}} + (0.1)\left(\frac{1}{2}\right) F_{1,k} = v_{1,k-\frac{1}{2}} + (0.05) F_{1,k} \qquad (1.19)$$

$$x_{1,k+1} = x_{1,k} + (0.1)\,v_{1,k+\frac{1}{2}}. \qquad (1.20)$$

From (1.16), then,

$$F_{1,0} = \left(-\frac{1}{1^3} + \frac{2}{1^6} \right)\left(-\frac{1}{1} \right) = -1,$$

so that (1.18) implies

$$v_{1,\frac{1}{2}} = -1.025.$$

In an analogous fashion, one finds that

$$v_{2,\frac{1}{2}} = v_{2,0} + (0.05)\,a_{2,0} \qquad \text{(Starter)}$$

$$v_{2,k+\frac{1}{2}} = v_{2,k-\frac{1}{2}} + (0.1)\,a_{2,k}, \quad k = 1,2,3,\ldots$$

$$x_{2,k+1} = x_{2,k} + (0.1)\,v_{2,k+\frac{1}{2}}, \quad k = 0,1,2,3\ldots,$$

and

$$v_{2,\frac{1}{2}} = 3 + (0.05)\frac{F_{2,0}}{1} = 3 + (0.05)F_{2,0}, \qquad \text{(Starter)} \qquad (1.21)$$

$$v_{2,k+\frac{1}{2}} = v_{2,k-\frac{1}{2}} + (0.1)\frac{F_{2,k}}{1} = v_{2,k-\frac{1}{2}} + (0.1)F_{2,k}, \quad k = 1, 2, 3, \ldots \quad (1.22)$$

$$x_{2,k+1} = x_{2,k} + (0.1)v_{2,k+\frac{1}{2}}, \qquad k = 0, 1, 2, \ldots. \quad (1.23)$$

Then, (1.16) and (1.21) yield

$$v_{2,\frac{1}{2}} = 3.05.$$

Since $v_{1,\frac{1}{2}}, v_{2,\frac{1}{2}}$ are now known, we can determine $x_{1,1}, x_{2,1}$ from (1.20) and (1.23) to yield

$$x_{1,1} = x_{1,0} + (0.1)v_{1,\frac{1}{2}} = -0.1025$$
$$x_{2,1} = x_{2,0} + (0.1)v_{2,\frac{1}{2}} = 1.305.$$

Next, knowing $x_{1,1}, x_{2,1}$, we can use (1.19) and (1.22) to determine $v_{1,\frac{3}{2}}, v_{2,\frac{3}{2}}$ as follows. For $k = 1$,

$$
\begin{aligned}
F_{1,1} &= \left(-\frac{1}{|x_{2,1}-x_{1,1}|^3} + \frac{2}{|x_{2,1} - x_{1,1}|^6} \right) \frac{x_{1,1} - x_{2,1}}{|x_{2,1} - x_{1,1}|} \\
&= \left(-\frac{1}{(1.305 + 0.1025)^3} + \frac{2}{(1.305 + 0.1025)^6} \right) \frac{-0.1025 - 1.305}{1.305 + 0.1025} \\
&= 0.101396
\end{aligned}
$$

$$F_{2,1} = -0.101396.$$

Thus, from (1.19),

$$v_{1,\frac{3}{2}} = v_{1,\frac{1}{2}} + (0.05)F_{1,1} = -1.019930,$$

while, from (1.22),

$$v_{2,\frac{3}{2}} = v_{2,\frac{1}{2}} + (0.1)F_{2,1} = 3.039860.$$

One next determines $x_{1,2}$ and $x_{2,2}$ from (1.20) and (1.23) to yield

$$x_{1,2} = x_{1,1} + (0.1)v_{1,\frac{3}{2}} = -0.204493.$$
$$x_{2,2} = x_{2,1} + (0.1)v_{2,\frac{3}{2}} = 1.608986.$$

The computation then continues in the indicated fashion until one has calculated for a fixed time period, which is usually prescribed by the constraints of the problem under consideration.

Note that in later discussions, confusion in notation will be avoided by using either h or Δt to represent a time step in a calculation.

Specific FORTRAN programs will be provided in later chapters. At present, however, a generic computer program for implementing the leap frog formulas is given as follows.

$ALGORITHM$ — PROGRAM LEAP FROG

Step 1. Set a time step h.

Step 2. Let distinct times t_k be given by $t_k = kh$, $k = 0, 1, 2, \ldots$

Step 3. Let $P(I)$ of mass $m(I)$ be N given bodies, $I = 1, 2, \ldots, N$.

Step 4. For each I, let $P(I)$ be initially at $x(I,0), y(I,0), z(I,0)$ with velocity components $vx(I,0), vy(I,0), vz(I,0)$.

Step 5. For each I, let the force on $P(I)$ at any time t be $(Fx(I,t), Fy(I,t), Fz(I,t))$ and let the acceleration be $(Ax(I,t), Ay(I,t), Az(I,t))$, with
$Ax(I,t) = Fx(I,t)/m(I)$
$Ay(I,t) = Fy(I,t)/m(I)$
$Az(I,t) = Fz(I,t)/m(I)$.

Step 6. For each I, apply the starter formulas
$vx(I,0.5) = vx(I,0) + \frac{1}{2}hAx(I,0)$
$vy(I,0.5) = vy(I,0) + \frac{1}{2}hAy(I,0)$
$vz(I,0.5) = vz(I,0) + \frac{1}{2}hAz(I,0)$

Step 7. For each I, determine positions and velocities sequentially by
$x(I,t_k + h) = x(I,t_k) + (h)vx(I,t_k + 0.5h), k = 0, 1, 2, \ldots$
$y(I,t_k + h) = y(I,t_k) + (h)vy(I,t_k + 0.5h), k = 0, 1, 2, \ldots$
$z(I,t_k + h) = z(I,t_k) + (h)vz(I,t_k + 0.5h), k = 0, 1, 2, \ldots$
$vx(I,t_p + 0.5h) = vx(I,t_p - 0.5h) + (h)Ax(I,t_p), p = k + 1.$
$vy(I,t_p + 0.5h) = vy(I,t_p - 0.5h) + (h)Ay(I,t_p), p = k + 1.$
$vz(I,t_p + 0.5h) = vz(I,t_p - 0.5h) + (h)Az(I,t_p), p = k + 1.$

Step 8. Stop when k = 999.

1.5.   Turbulence

Turbulence is the most common, yet least understood, form of fluid motion. Let us the first describe some related concepts, views, and results from the points of view of engineering, fluid theory, and numerical analysis. As will be clear from the discussion, overlapping between any two of these areas is relatively common.

1.5.1. *Engineering*

From an engineering point of view, turbulent flow may be characterized qualitatively as fluid motion displaying seemingly random behavior. More precisely, if a fluid flow is one realization of repeatable experiment, then the flow is called turbulent if the velocity field, either in whole or in part, changes to a degree greater than experimental error each time the experiment is performed. Even though the flow is deterministic, turbulence results because the flow is acutely sensitive to minute, uncontrollable differences in boundary and/or initial data, whose effects are amplified in time to a measurable scale. *The flow is characterized also by the random appearance and disappearance of many vortices.* However, even though understanding the dynamics of these vortical structures is considered to be an important part of current turbulence research, many related physical and mathematical issues are unresolved. (Bernard (1998), Speziale and So (1998)).

Also, it is known that *a strong current across a customary laminar flow initiates turbulent flow* (Schlichting (1960)).

It is also thought by some (Yakhot and Orszag (1986)) that sub continuum perturbations, such as those responsible for Brownian motion, may initiate instabilities which lead to turbulence.

1.5.2. *Theoretical*

A fundamental pure number, called the Reynolds number Re, is defined as follows:

$$Re = UL/\nu,$$

in which U and L are reference velocity and length scales and ν is the kinematic viscosity of the fluid under consideration, is assumed to be related to turbulence. It is believed that Re large results in turbulent flow. Note that the Reynolds number is not related to any assumed equations of fluid flow.

As for the overall theoretical modelling of turbulent flow, several major approaches have been formulated and explored. Beginning with Taylor's seminal paper (Taylor (1921)), one school of study (Favre (1964)) has emphasized the statistical approach. A second major school of thought, following fundamental papers of Landau (1944), Hopf (1948), and Ruelle and Takens (1971) uses Galerkin approximations to simplify the Navier–Stokes equations and bifurcation theory to analyze the resulting ordinary differential system (Barenblatt, Iooss, and Joseph (1983)). A still third approach utilizes statistical thermodynamics (Chorin (1994), Malkus (1960)). Unfortunately, important and realistic aspects of turbulent motions have defied inclusion in all the models just described (Barenblatt, Iooss, and Joseph (1983), Markatos (1986), Zabusky (1968)). Thus, for example, whereas homogeneous turbulence has received intensive theoretical study (Batchelor (1960)), it is not known to exist anywhere in Nature.

1.5.3. *Numerical*

The dominant numerical approaches to the simulation of turbulent flow have centered on computational solution of the Navier–Stokes equations for large Reynolds numbers. This has been shown to be invalid if one uses the classical (non averaged) Navier–Stokes equations. In the words of Ladyzhenskaya (1969): "The results given in this book support the belief that it is reasonable to use the Navier–Stokes equations to describe the motions of a viscous fluid in the case of Reynolds numbers which do not exceed certain limits. They partially refute the statements described above concerning the properties of solutions of the Navier–Stokes equations, and they force us to find other explanations for observed phenomena in real fluids, in particular, for the familiar paradoxes involving viscous fluids. Apparently, in seeking these explanations, one must ignore the fact that if a large force **f** acts on the fluid for an extended interval of time, then the quantities $D_x^m v_k$ (where $\mathbf{v} = (v_1, v_2, v_3)$ is the solution) can become so large that the assumption that they are comparatively small, made in deriving the Navier–Stokes equations from the statistical Maxwell–Boltzmann equations, will no longer be satisfied, just as other assumptions of the Stokes theory, i.e. the assumption that the kinematic viscosity and the thermal regime are constant, will be far from valid. Because of this, it is hardly possible to explain the transition from laminar to turbulent flows within the framework of the classical Navier–Stokes theory."

Replacement of a numerical solution of the classical Navier–Stokes equations by a numerical solution of averaged fluid equations using large Reynolds numbers leave many related mathematical issues unresolved and have also led to results which are contradicted by experiment (Bernard (1998)).

1.6. Overview

In order not to spread our attention too broadly, we will concentrate, primarily, in both two and three dimensions on a prototype fluid flow called *cavity flow.* For such problems we will use two significant markers discussed in Section 1.5, namely, that turbulent flow follows when a strong cross-current develops and imposes itself on a laminar flow (Schlichting (1960)), and that turbulent flow exhibits the rapid appearance and disappearance of many vortices (Kolmogorov (1964)). However, initially, we will work in the spirit of nano physics and will proceed into the large only in a later chapter, Chapter 6. On the molecular level, we may find somewhat different results at first that one might expect in the large. Indeed, researchers in both nano physics and nano technology are finding that this is exactly the case. However, our results will confirm the conjecture of Yakhot and Orszag (1986) that sub continuum perturbations, such as those responsible for Brownian motion, do initiate instabilities which lead to turbulence. Indeed, by working first on the molecular level we will show clearly the mechanisms by which turbulence develops.

Chapter 2

Molecular Cavity Flow of Argon Vapor in Two Dimensions

2.1. Introduction

A Lennard–Jones potential for argon is (Hirschfelder, Curtiss and Bird (1967)):

$$\phi(r_{ij}) = (6.848)10^{-14} \left[\frac{3.418^{12}}{r_{ij}^{12}} - \frac{3.418^6}{r_{ij}^6} \right] \text{ erg } \left(\frac{\text{grcm}^2}{\text{sec}^2} \right) \qquad (2.1)$$

in which r_{ij} is measured in angstroms (Å). The force $\vec{F}_{ij}$ exerted on P_i by P_j is

$$\vec{F}_{ij} = (6.848)10^{-6} \left[\frac{12(3.418)^{12}}{r_{ij}^{13}} - \frac{6(3.418)^6}{r_{ij}^7} \right] \frac{\vec{r}_{ji}}{r_{ij}} \text{ dynes } \left(\frac{\text{grcm}}{\text{sec}^2} \right) \qquad (2.2)$$

which reduces readily to

$$\vec{F}_{ij} = \left[\frac{209.0}{r_{ij}^{13}} - \frac{0.06551}{r_{ij}^7} \right] \frac{\vec{r}_{ji}}{r_{ij}} \text{ dynes } \left(\frac{\text{grcm}}{\text{sec}^2} \right). \qquad (2.3)$$

Note also that

$$F_{ij} = \|\vec{F}_{ij}\| = \left[\frac{209.0}{r_{ij}^{13}} - \frac{0.06551}{r_{ij}^7} \right], \qquad (2.4)$$

so that $F_{ij}(r_{ij}) = 0$ implies that $r_{ij} = 3.837\,\text{Å}$, which is the *equilibrium* distance.

2.2. Equations of Motion for Argon Vapor

From the discussion in Section 2.1, it follows that the equation of motion for a single argon vapor atom P_i acted on by a single argon vapor atom $P_j, i \neq j$, is

$$m\vec{a}_i = \left[\frac{209.0}{r_{ij}^{13}} - \frac{0.06551}{r_{ij}^{7}} \right] \frac{\vec{R}_{ji}}{R_{ij}} \left(\frac{\text{grcm}}{\text{sec}^2} \right). \tag{2.5}$$

Since the mass of argon atom is $(6.63)10^{-23}$ gr, and since $r = 10^8 R$, the latter equation is equivalent to

$$\vec{a}_i = \frac{10^{23}}{6.63} \left[\frac{209.0}{r_{ij}^{13}} - \frac{0.06551}{r_{ij}^{7}} \right] \frac{\vec{r}_{ji}}{r_{ij}} \left(\frac{\text{cm}}{\text{sec}^2} \right).$$

Replacing centimeters by angstroms and seconds by picoseconds then yields readily

$$\vec{a}_i = 10^6 \left[\frac{315.2}{r_{ij}^{13}} - \frac{0.09881}{r_{ij}^{7}} \right] \frac{\vec{r}_{ji}}{r_{ij}} \left(\frac{\text{Å}}{\text{ps}^2} \right) \tag{2.6}$$

or

$$\frac{d^2\vec{r}_i}{dT^2} = 98810 \left[\frac{3190}{r_{ij}^{13}} - \frac{1}{r_{ij}^{7}} \right] \frac{\vec{r}_{ji}}{r_{ij}} \left(\frac{\text{Å}}{\text{ps}^2} \right) \tag{2.7}$$

which is the equation we will use.

On the molecular level, however, the effective force on P_i is *local*, that is it is determined only by close molecules. For argon atoms we choose the local interaction distance to be $D = 2.5\sigma = 2.5(3.418) = 8.545$ Å.

From (2.7), then, the dynamical equation for argon vapor atom P_i will be

$$\frac{d^2\vec{r}_i}{dT^2} = 98810 \left[\frac{3190}{r_{ij}^{13}} - \frac{1}{r_{ij}^{7}} \right] \frac{\vec{r}_{ji}}{r_{ij}}; \quad r_{ij} < D. \tag{2.8}$$

The equations of motion for a system of N argon vapor atoms are then

$$\frac{d^2\vec{r}_i}{dT^2} = 98810 \sum_{\substack{j \\ j \neq i}} \left[\frac{3190}{r_{ij}^{13}} - \frac{1}{r_{ij}^{7}} \right] \frac{\vec{r}_{ji}}{r_{ij}}; \quad i = 1, 2, 3, \ldots, N; \ r_{ij} < D. \tag{2.9}$$

Observe that on the molecular level gravity can be neglected since $980 \, \text{cm/sec}^2 = (980)10^{-16} \text{Å/ps}^2$.

2.3. The Cavity Problem

In two dimensions consider the square of side l shown in Fig. 2.1.
The interior is called the cavity or the basin of the square. The sides
AB, BC, CD, DA are called the walls. Let argon vapor fill the basin.
The top wall CD, alone, is allowed to move. It moves in the X direction at a constant speed V, called the *wallspeed*. Also it is allowed an
extended length so that the fluid is always completely enclosed by four
walls. (For the reasonableness of this assumption, see the treadmill appa-
ratus of Koseff and Street (1994)). Then the cavity problem is to describe
the gross motion of the fluid for various choices of V, which will be given
in Å/ps.

Consider the cavity problem with $l = 230.22$ Å. Note that for a regular
triangle with edge length 3.837 Å, which is the equilibrium distance, the
altitude is 3.323 Å. Using these values we now construct a regular triangular

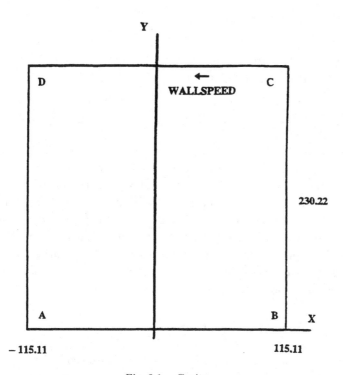

Fig. 2.1. Cavity

Fig. 2.2.   4235 atoms.

grid of 4235 points on the basin as follows:

$$x(1) = -115.11, \quad y(1) = 0$$
$$x(i) = x(i-1) + 3.837, \quad y(i) = 0, \qquad i = 2, 61$$
$$x(62) = -113.1915, \quad y(62) = 3.323$$
$$x(i) = x(i-1) + 3.837, \quad y(i) = 3.323, \qquad i = 63, 121$$
$$x(i) = x(i-121), \quad y(i) = y(i-121) + 6.646, \qquad i = 122, 4235.$$

At each point $(x(i), y(i))$ we set an argon atom $P_i, i = 1, 4235$. This array is shown in Fig. 2.2 and provides the initial positions of the atoms.

To complete the initial data we need the initial velocities of the atoms. Let us choose the temperature to be $35°$ C, so that $v = 3.58\,\text{Å/ps}$. Each atom is then assigned a speed of $3.58\,\text{Å/ps}$ in either the X or Y direction, determined at random, with its sign $(\pm)$ also determined at random. Once the wallspeed V is prescribed, we are ready to solve the resulting cavity problem using the leap frog formulas for Eq. (2.9) with $N = 4235$. Note that the number of equations is 8470.

2.4.   Computational Considerations

For time step Δt (ps), and $t_k = k\Delta t$, two problems must be considered relative to the computations. The first problem is to prescribe a protocol when, computationally, an atom has crossed a wall into the exterior of the cavity. For each of the lower three walls, we will proceed as follows (no slip condition). The position will be reflected back symmetrically, relative to the wall, into the interior of the basin, the velocity component tangent to the wall will be set to zero and the velocity component perpendicular to the wall will be multiplied by -1. If the atom has crossed the moving wall, then its position will be reflected back symmetrically, its Y component of velocity will be multiplied by -1, and its X component of velocity will be increased by the wallspeed V.

The second problem derives from the fact that an instantaneous velocity field for molecular motion is Brownian. In order to better interpret gross fluid motion , we will introduce average velocities as follows. For J a positive integer, let particle P_i be at $(x(i, k), y(i, k))$ at t_k and at $(x(i, k - J), y(i, k - J))$ at t_{k-J}. Then the average velocity $\vec{v}_{i,k,J}$ of P_i at t_k is defined by

$$\vec{v}_{i,k,J} = \left(\frac{x(i, k) - x(i, k - J)}{J \Delta t}, \frac{y(i, k) - y(i, k - J)}{J \Delta t} \right). \qquad (2.10)$$

In the examples to be described, we will discuss results for various values of J.

Observe also that the connection between the wall speed V and the Reynolds number Re is given by (Pan and Acrivos (1967))

$$Re = |V| B / \nu, \qquad (2.11)$$

in which B is the span, that is, the length CD in Fig. 2.1, and ν is the average kinematic viscosity.

2.5.   Examples of Primary Vortex Generation

Consider first the parameter choices $V = -50, J = 25000, \Delta t = 0.00002$. Figures 2.3–2.6 show the development of a primary vortex at the respective times $T = 0.5, 1.0, 2.0, 3.5$. An initial compression wave is evident in Fig. 2.3. The resulting compression in the lower right corner results in an upward repulsive effect which initiates the development of the counterclockwise vortex shown in Fig. 2.6. The mean of the squares of the speeds at $T = 3.5$ is 156 Å/ps. Other values of J which were studied were 20000, 15000, and 10000, each of which yielded results similar to those of Figs. 2.3–2.6.

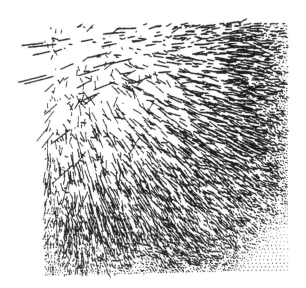

Fig. 2.3. $V = -50$, $J = 25000$, $T = 0.5$.

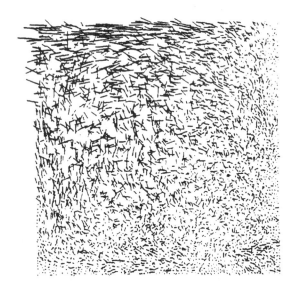

Fig. 2.4. $V = -50$, $J = 25000$, $T = 1.0$.

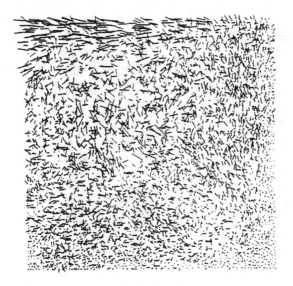

Fig. 2.5. $V = -50$, $J = 25000$, $T = 2.0$.

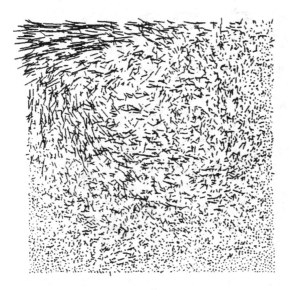

Fig. 2.6. $V = -50$, $J = 25000$, $T = 3.5$.

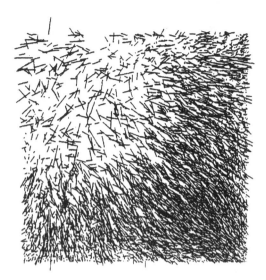

Fig. 2.7. $V = -100$, $J = 15000$, $T = 0.5$.

Fig. 2.8. $V = -100$, $J = 15000$, $T = 1.5$.

For the parameter choices $V = -100, J = 15000, \Delta t = 0.00002$, Figs. 2.7–2.10 show the development of a primary vortex at the respective times $t = 0.5, 1.5, 2.0, 2.5$. An initial compression wave is seen in Fig. 2.7 and again the compression in the lower right corner results in an upward

Fig. 2.9. $V = -100$, $J = 15000$, $T = 2.0$.

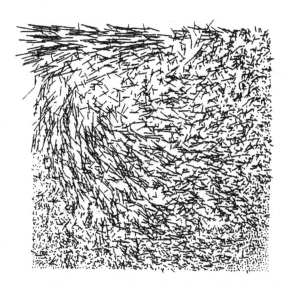

Fig. 2.10. $V = -100$, $J = 15000$, $T = 2.5$.

repulsive effect which yield the counterclockwise vortex in Fig. 2.10. The vortex in Fig. 2.10 is larger than that in Fig. 2.6 and has developed more quickly. The mean of the squares of the speeds at $T = 2.5$ is $222\,\text{Å/ps}$. Other values of J which were studied were 25000, 20000, and 10000, each of which yielded results similar to those of Figs. 2.7–2.10.

Other choices of V were -25 and -250, which yielded the expected results. For $V = -25$, the resulting primary vortex was smaller than that in Fig. 5.4. For $V = -250$, the resulting primary vortex was larger than that in Fig. 5.8.

2.6.   Example of Turbulent Flow

Recall now that, *in the large*, turbulent flows have two well defined criteria:

(1) A strong current develops across the usual primary vortex direction (Kolmogorov (1964)), and

(2) Many small vortices appear and disappear quickly (Schlichting (1960)).

We will examine these criteria in our molecular calculations.

We take the following approach to generating turbulent flows. For a sufficiently large magnitude of the wallspeed V, let us show that turbulence results when, for given Δt, a stable calculation results but no J exists which yields a primary vortex.

Let us then set $V = -3000, \Delta t = 4(10)^{-7}$. The motion was simulated to $T = 1.11$. Typical results are shown at $T = 0.03, 0.09, 0.18, 0.27, 0.45, 0.72$, 1.08, and 1.11 in Figs. 2.11–2.18, respectively. Figure 2.11 shows the very rapid development of a compression wave. Figures 2.12–2.14 show what appears to be a large primary vortex. However, Figs. 2.15–2.18 show the development of a strong vertical current in the left portion of the figure. Figure 2.18 shows that this current runs across the usual primary vortex direction not only in the left portion of the figure but, to a lesser degree, also in the right portion. Thus, Criterion (1) is valid.

We next define the concept of a small vortex. For $3 \leq M \leq 6$, we define a *small vortex* as a flow in which M molecules nearest to an $(M+1)$st molecule rotate either clockwise or counterclockwise about the $(M + 1)$st molecule and, in addition, the $(M + 1)$st molecule lies *interior* to a simple polygon determined by the given M molecules. With this definition, Fig. 2.19 shows that the flow in Fig. 2.17 has 198 small vortices at $T = 1.08$, while

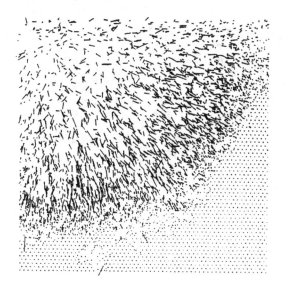

Fig. 2.11. $V = -3000$, $J = 75000$, $T = 0.03$.

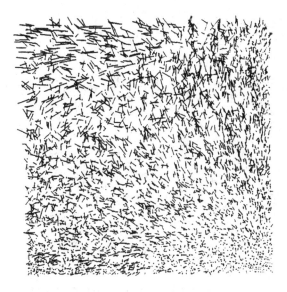

Fig. 2.12. $V = -3000$, $J = 75000$, $T = 0.09$.

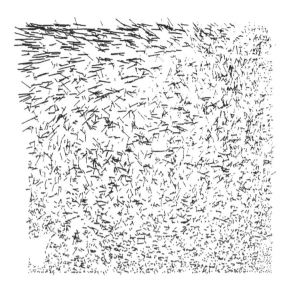

Fig. 2.13. $V = -3000$, $J = 75000$, $T = 0.18$.

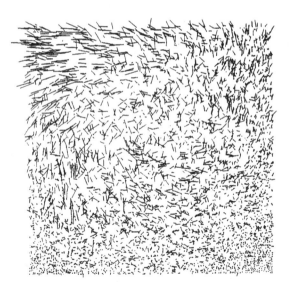

Fig. 2.14. $V = -3000$, $J = 75000$, $T = 0.27$.

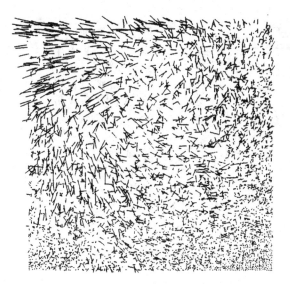

Fig. 2.15. $V = -3000$, $J = 75000$, $T = 0.45$.

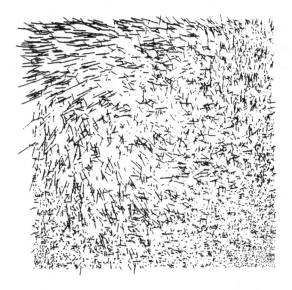

Fig. 2.16. $V = -3000$, $J = 75000$, $T = 0.72$.

Fig. 2.17. $V = -3000, \; J = 75000, \; T = 1.08.$

Fig. 2.18. $V = -3000, \; J = 75000, \; T = 1.11.$

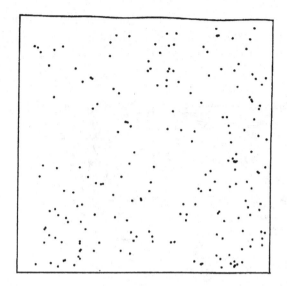

Fig. 2.19.   198 small vortices in Fig. 2.17.

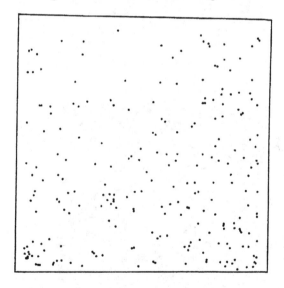

Fig. 2.20.   236 small vortices in Fig. 2.18.

Fig. 2.20 shows that at $T = 1.11$, only 0.03 picoseconds later, the resulting flow has 236 small vortices which are completely different than those in Fig. 2.19.

Figures 2.21–2.23 show the same general flow as in Fig. 2.18, but using $J = 60000$, 45000, and 30000, respectively, so that the choice of J is immaterial to the development of turbulence.

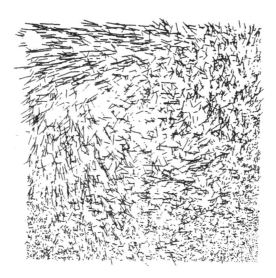

Fig. 2.21. $V = -3000$, $J = 60000$, $T = 1.11$.

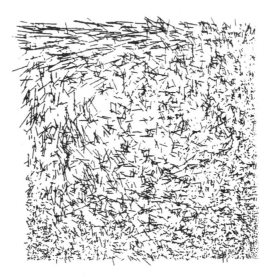

Fig. 2.22. $V = -3000$, $J = 45000$, $T = 1.11$.

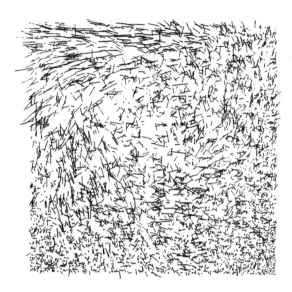

Fig. 2.23. $V = -3000$, $J = 30000$, $T = 1.11$.

2.7. Remark

Recall now that it is known that nano mechanical results may differ from those in the large. This is indeed exactly the case in our computations, for Criterion 2 is also valid for primary vortex motions, like those shown in Figs. 2.3–2.10.

2.8. The **FORTRAN PROGRAM ARGON.FOR**

A typical computer program for the examples in this chapter is now given.

FORTRAN PROGRAM ARGON.FOR

```
C  TOTAL NUMBER OF PARTICLES IS  4235.
c P AND Q ARE MOLECULAR PARAMETERS 7 AND 13.
     double precision XO(4235),YO(4235),VXO(4235),VYO(4235),
     1X(4235,3),Y(4235,3),VX(4235,2),VY(4235,2),
     1ACX(4235),ACY(4235),r2(4235),r4(4235),fx(4235),
     1fy(4235),f(4235)
     1,xke,R6(4235),R8(4235)
```

```
      OPEN (UNIT=21,file='argon.dat',status='old')
      OPEN (UNIT=31,file='argon.out',status='new')
      OPEN (UNIT=71,file='argon.out1',status='new')
      OPEN (UNIT=72,file='argon.out2',status='new')
      OPEN (UNIT=73,file='argon.out3',status='new')
      OPEN (UNIT=74,file='argon.out4',status='new')
      OPEN (UNIT=75,file='argon.out5',status='new')
      OPEN (UNIT=76,file='argon.out6',status='new')
      K=1
      kk=1
      KPRINT=50000
      wallspeed= -3000.
      READ (21,10) (XO(I),YO(I),VXO(I),VYO(I),I=1,4235)
10    FORMAT (2f12.3,2f12.3)
      DO 30 I=1,4235
      X(I,1)=XO(I)
      Y(I,1)=YO(I)
      VX(I,1)=VXO(I)
      VY(I,1)=VYO(I)
30    CONTINUE
      GO TO 3456
C Cycle.
65    DO 70 I=1,4235
      X(I,1)=X(I,2)
      Y(I,1)=Y(I,2)
      VX(I,1)=VX(I,2)
      VY(I,1)=VY(I,2)
70    CONTINUE
      DO 701 I=1,4235
      ACX(I)=0.
      ACY(I)=0.
      F(I)=0.
701   CONTINUE
3456  DO 78 I=1,4234
      ACXI=ACX(I)
      ACYI=ACY(I)
      XI=X(I,1)
      YI=Y(I,1)
      IP1=I+1
```

```
      DO 77 J=IP1,4235
      R2(J)=(XI-X(J,1))**2+(YI-Y(J,1))**2
      R6(J)=R2(J)*R2(J)*R2(J)
      R8(J)=R6(J)*R2(J)
C in this program the local interaction distance is 2.5sigma.
c sigma = 3.418
C Force calculation
      if (r2(j).gt.73.) go to 9000
      F(J)=98810.*(-1.+3190/(R6(J)))
     1   /(R8(J))
      go to 9001
9000  f(j)=0.0
9001  ACX(J)=ACX(J)-F(J)*(XI-X(J,1))
      ACY(J)=ACY(J)-F(J)*(YI-Y(J,1))
      ACXI=ACXI+F(J)*(XI-X(J,1))
      ACYI=ACYI+F(J)*(YI-Y(J,1))
77    CONTINUE
      ACX(I)=ACXI
      ACY(I)=ACYI
78     CONTINUE
c Note, gravity is no consequence on the molecular level.
      DO 7123 I=1,4235
      VX(I,2)=VX(I,1)+0.0000004*ACX(I)
      VY(I,2)=VY(I,1)+0.0000004*ACY(I)
      X(I,2)=X(I,1)+0.0000004*VX(I,2)
      Y(I,2)=Y(I,1)+0.0000004*VY(I,2)
7123     CONTINUE
c at this point we insert wall reflection.
      do 995 i=1,4235
      if (x(i,2).lt.115.11) go to 992
      x(i,2)=230.22-x(i,2)
      vx(i,2)=-1.0*vx(i,2)
      vy(i,2)=0.0*vy(i,2)
992   if (x(i,2).gt.-115.11) go to 993
      x(i,2)=-230.22-x(i,2)
      vx(i,2)=-1.0*vx(i,2)
      vy(i,2)=0.0*vy(i,2)
993   if (y(i,2).gt.0.0) go to 994
      y(i,2)=-y(i,2)
```

```
      vx(i,2)=0.0*vx(i,2)
      vy(i,2)=-1.00*vy(i,2)
994   if (y(i,2).lt.229.287) go to 995
      y(i,2)=458.574-y(i,2)
      vx(i,2)=vx(i,2)+wallspeed
      vy(i,2)=-1.*vy(i,2)
995   continue
      K=K+1
      if (k.eq.2) go to 500
      go to 501
500   write (71,10) (x(i,2),y(i,2),vx(i,2),vy(i,2),i=1,4235)
501   If (k.eq.15000) go to 502
      go to 503
502   write (72,10) (x(i,2),y(i,2),vx(i,2),vy(i,2),i=1,4235)
503   If (k.eq.30000) go to 504
      go to 505
504   write (73,10) (x(i,2),y(i,2),vx(i,2),vy(i,2),i=1,4235)
505   If (k.eq.45000) go to 506
      go to 507
506   write (74,10) (x(i,2),y(i,2),vx(i,2),vy(i,2),i=1,4235)
507   if (k.eq.60000) go to 508
      go to 509
508   write (75,10) (x(i,2),y(i,2),vx(i,2),vy(i,2),i=1,4235)
509   if (k.eq.75000) go to 510
      go to 511
510   write (76,10) (x(i,2),y(i,2),vx(i,2),vy(i,2),i=1,4235)
511   continue
82    if (k.lt.75000) go to 65
      STOP
      END
```

2.9.   The Fortran Program **YOUWIN.FOR**

A Fortran program for determining small vortices is given as follows.

FORTRAN 90 PROGRAM YOUWIN.FOR

```
C THIS PROGRAM GIVES ALL POSSIBLE SMALL VORTICES IN FIG.DAT 4.
C IT USES TWO RESULTS RELATED TO THE AREA OF A TRIANGLE WHEN
C THE AREA IS GIVEN IN DETERMINANT FORM:
```

```
C (1) IF (x1,y1),(x2,y2),(x3,y3) ARE ORIENTED COUNTERCLOCKWISE, THE
C AREA IS POSITIVE. OTHERWISE, IT IS NEGATIVE.
C (2) A POINT P INSIDE THE TRIANGLE SATISFIES THE CONDITION THAT
C THE SUM OF THE ABSOLUTE VALUES OF THE THREE AREAS IT MAKES
C WITH EACH PAIR OF VERTICES IS LESS THAN OR EQUAL TO THE
C ABSOLUTE VALUE OF THE AREA OF THE TRIANGLE.

C THE PROGRAM WAS WRITTEN BY V. CASULLI.

        DOUBLE PRECISION R2(4235),x(4235),y(4235),vx(4235),vy(4235)
       1,vxx(4235),vyy(4235)
          DIMENSION xv(0:99),yv(0:99),uv(99),vv(99),rr(99)
        OPEN (UNIT=21,FILE='fig.dat4',STATUS='old')
        OPEN (UNIT=22,FILE='vortex.out',STATUS='unknown')
        OPEN (UNIT=23,FILE='vortex.all',STATUS='unknown')
        READ (21,10) (x(i),y(i),vx(i),vy(i),i=1,4235)
10      FORMAT (4f12.6)
11      FORMAT (4f12.6,f10.6)
12      FORMAT (I10)
        DO 1112 i=1,4235
        kx=0
          xv(0)=x(i)
          yv(0)=y(i)
        DO 1111 j=1,4235
          IF(i.eq.j)GO TO 1111
          R2(j)=(x(i)-x(j))**2+(y(i)-y(j))**2
c WE USE A DISTANCE PARAMETER TO INSURE THE VORTEX IS SMALL.
          IF(r2(j).gt.16.0) GO TO 1111
          vxx(j)=vx(j)-vx(i)
          vyy(j)=vy(j)-vy(i)
          kx=kx+1
          xv(kx)=x(j)
          yv(kx)=y(j)
          uv(kx)=vxx(j)
          vv(kx)=vyy(j)
          rr(kx)=r2(j)
        WRITE (22,11) x(j),y(j),vxx(j),vyy(j),r2(j)
1111    CONTINUE
        WRITE (22,12) kx
        IF(kx.lt.3)GO TO 1112

        CALLorder(kx,xv,yv,uv,vv,rr)
        CALL vortex(kx,xv,yv,uv,vv)

1112 CONTINUE
     STOP
     END

        SUBROUTINE order(kx,xv,yv,uv,vv,rr)
        DIMENSION xv(0:99),yv(0:99),uv(99),vv(99),rr(99)
        DO k=1,kx-1
        DO j=k+1,kx
          IF(rr(k).gt.rr(j))THEN
```

```
          temp=xv(k)
          xv(k)=xv(j)
          xv(j)=temp
          temp=yv(k)
          yv(k)=yv(j)
          yv(j)=temp
          temp=uv(k)
          uv(k)=uv(j)
          uv(j)=temp
          temp=vv(k)
          vv(k)=vv(j)
          vv(j)=temp
          temp=rr(k)
          rr(k)=rr(j)
          rr(j)=temp
          END IF
          END DO
          END DO
          RETURN
          END


          SUBROUTINE vortex(kx,x,y,u,v)
          DIMENSION x(0:99),y(0:99),u(99),v(99),a(99)
          kk=2
          DO k=3,kx
          DO m=1,k-1
            at=abs(area(x(m),y(m),x(m+1),y(m+1),x(k),y(k)))
            a1=abs(area(x(0),y(0),x(m+1),y(m+1),x(k),y(k)))
            a2=abs(area(x(m),y(m),x(0  ),y(0  ),x(k),y(k)))
            a3=abs(area(x(m),y(m),x(m+1),y(m+1),x(0),y(0)))
            IF(a1+a2+a3.le.at)THEN
              kk=k
              GO TO 10
            END IF
          END DO
          END DO
10        CONTINUE
          IF(kk.le.2)RETURN


          DO i=1,kk
          a(i)=area(x(0),y(0),x(i),y(i),x(i)+u(i),y(i)+v(i))
          END DO
            DO i=2,kk
            IF(a(1)*a(i).le.0.0)RETURN
            END DO
          TYPE *,'You win! Particle',x(0),y(0), 'is center of a vortex.'
          WRITE(23,'(2f12.6)')x(0),y(0)
```

```
RETURN
END

FUNCTION AREA(x1,y1,x2,y2,x3,y3)
AREA=x1*y2+x2*y3+x3*y1-x3*y2-x2*y1-x1*y3
RETURN
END
```

Chapter 3

Molecular Cavity Flow of Air Vapor in Two Dimensions

3.1. Molecular Formulas

It is rather interesting that even though air is heterogeneous and consists of a variety of atoms and molecules, experimental Lennard–Jones potentials are readily available only for homogeneous air (Hirschfelder, Curtiss and Bird [2]). One such potential is

$$\phi(r_{ij}) = (5.36)10^{-14} \left[\frac{3.617^{12}}{r_{ij}^{12}} - \frac{3.617^6}{r_{ij}^6} \right] \text{ erg}, \qquad (3.1)$$

in which r_{ij} is measured in angstroms (Å). The force $\vec{F}_{ij}$ exerted on P_i by P_j is then

$$\vec{F}_{ij} = (5.36)10^{-6} \left[\frac{12(3.617^{12})}{r_{ij}^{13}} - \frac{6(3.617^6)}{r_{ij}^7} \right] \frac{\vec{r}_{ji}}{r_{ij}} \text{ dynes} \qquad (3.2)$$

and the equilibrium distance is $r_{ij} = 4.06\,\text{Å}$. On the molecular level, the effective force on P_i is local, so that only molecules within a distance $D = 2.5\sigma = 2.5(3.617)\,\text{Å}$ are considered.

Before proceeding to dynamical considerations, it is necessary to characterize carefully the hypothetical air molecule to be used. We assume that the air to be used is *non dilute* and *dry*. Dry air (Masterton and Slowinski [3]) consists primarily of 78% N_2, 21% O_2, and 1% Ar, whose respective masses are

$$m(N_2) = 28(1.660)10^{-24}\,\text{gr}$$
$$m(O_2) = 32(1.660)10^{-24}\,\text{gr}$$
$$m(Ar) = 40(1.660)10^{-24}\,\text{gr}.$$

We now characterize an "air" molecule A as consisting of proportionate amounts of N_2, O_2, and Ar and having mass

$$m(A) = [0.78(28) + 0.21(32) + 0.01(40)](1.660)10^{-24} = (4.807)10^{-23}\,\text{gr}.$$
$$(3.3)$$

From (3.2) and (3.3) it follows that the acceleration, in $\text{Å}/(\text{ps})^2$, of an air molecule P_i due to interaction with an air molecule P_j satisfies the equation

$$\vec{a}_i = (149795)\left[\frac{4478}{r_{ij}^{13}} - \frac{1}{r_{ij}^7}\right]\frac{\vec{r}_{ji}}{r_{ij}}\ \left(\frac{\text{Å}}{\text{ps}^2}\right);\quad r_{ij} < D. \qquad (3.4)$$

The equations of motion for a system of air molecules are then

$$\frac{d^2\vec{r}_i}{dT^2} = (149795)\sum_{\substack{j \\ j \neq i}}\left[\frac{4478}{r_{ij}^{13}} - \frac{1}{r_{ij}^7}\right]\frac{\vec{r}_{ji}}{r_{ij}}\ \left(\frac{\text{Å}}{\text{ps}^2}\right);$$

$$i = 1, 2, 3, \ldots, N;\quad r_{ij} < D. \quad (3.5)$$

3.2. The Cavity Problem

We consider again a cavity problem. In two dimensions consider the square of side 243.6 Å shown in Fig. 3.1. The interior is the cavity or the basin. The sides are the walls. Let the basin be filled with air. The top wall, alone, is allowed to move in the X direction with wallspeed V. The cavity problem is to describe the gross motion of the fluid for various choices of V, which will be given in Å/ps.

3.3. Initial Data

Note that for a regular triangle with edge length 4.06 Å, which is the equilibrium distance, the altitude is 3.516 Å. Using these values we now construct a regular triangular grid of 4235 points on the basin as follows:

$$x(1) = -121.8,\quad y(1) = 0$$
$$x(i) = x(i-1) + 4.06,\quad y(i) = 0,\quad i = 2, 61$$
$$x(62) = -119.77,\quad y(62) = 3.516$$
$$x(i) = x(i-1) + 4.06,\quad y(i) = 3.516,\quad i = 63, 121$$
$$x(i) = x(i-121),\quad y(i) = y(i-121) + 7.032,\quad i = 122, 4235.$$

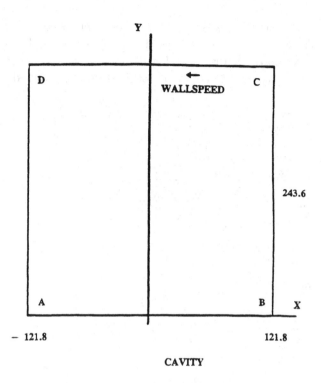

Fig. 3.1.   Cavity.

At each point $(x(i), y(i))$ we set an air molecule P_i, $i = 1, 4235$. This array is that shown in Fig. 3.2 and provides the initial positions of the molecules.

To complete the initial data we need the initial velocities of the molecules. In two dimensions at 35°C the speed of an air molecule is 4.207 Å/ps. Each molecule is then assigned a speed of 4.207 Å/ps in either the X or Y direction, determined at random, with its sign ($\pm$) also determined at random. Once the wallspeed V is prescribed, we are ready to solve the resulting cavity problem using the leap frog formulas for Eq. (3.5) with $N = 4235$.

The computational considerations are those given in Section 2.4.

3.4.   Examples of Primary Vortex Generation

Consider first the parameter choices $V = -50, J = 25000, \Delta t = 0.00002$. Figures 3.3–3.6 show the development of a primary vortex at the respective times $T = 0.5, 1.0, 2.0, 4.0$. An initial compression wave is evident in Fig. 3.3. The resulting compression in the lower right corner results in an upward repulsive effect which initiates the development of the counterclockwise vortex shown in Fig. 3.6. The mean of the squares of the speeds at $T = 4.0$ is $120\,\text{Å/ps}$. Other values of J which were studied were

Fig. 3.2. 4235 molecules.

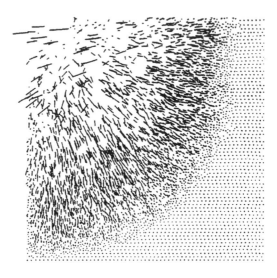

Fig. 3.3. $V = -50$, $J = 25000$, $T = 0.5$.

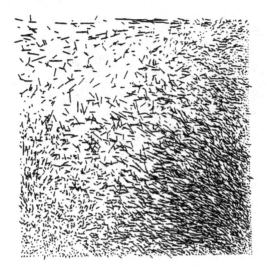

Fig. 3.4. $V = -50,\ J = 25000,\ T = 1.0$.

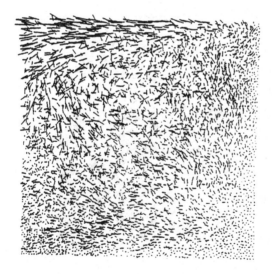

Fig. 3.5. $V = -50,\ J = 25000,\ T = 2.0$.

20000, 15000, and 10000, each of which yielded results similar to those of Figs. 3.3–3.6.

For the parameter choices $V = -100, J = 20000, \Delta t = 0.00002$, Figs. 3.7–3.10 show the development of a primary vortex at the respective

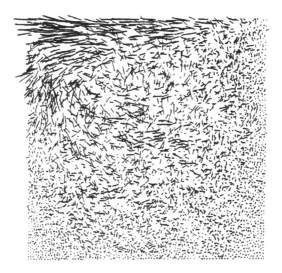

Fig. 3.6. $V = -50$, $J = 25000$, $T = 4.0$.

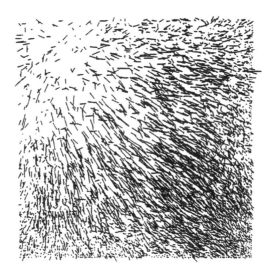

Fig. 3.7. $V = -100$, $J = 20000$, $T = 0.5$.

times $t = 0.5$, 1.0, 2.0, 3.0. An initial compression wave is seen in Fig. 3.7 and again the compression in the lower right corner results in an upward repulsive effect which results in the counterclockwise vortex in Fig. 3.10. The vortex in Fig. 3.10 is larger than that in Fig. 3.6 and has developed

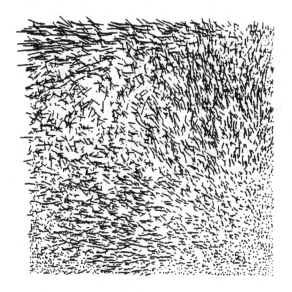

Fig. 3.8. $V = -100$, $J = 20000$, $T = 1.0$.

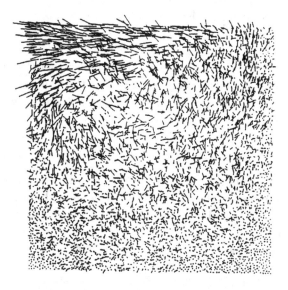

Fig. 3.9. $V = -100$, $J = 20000$, $T = 2.0$.

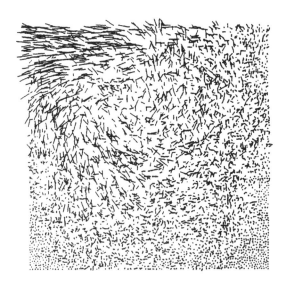

Fig. 3.10. $V = -100,\ J = 20000,\ T = 3.0.$

more quickly. The mean of the squares of the speeds at $T = 3.0$ is $234\ \text{Å/ps}$. Other values of J which were studied were 25000, 15000, and 10000, each of which yielded results similar to those of Figs. 3.7–3.10.

Other choices of V were -25 and -250, which yielded the expected results. For $V = -25$ the resulting primary vortex was smaller than that in Fig. 3.6 and the mean speed was lower. For $V = -250$, the resulting primary vortex was larger than that in Fig. 3.10 and the mean speed was greater.

3.5. Turbulent Flow

We again take the following approach to generating turbulent flows. For a sufficiently large magnitude of the wallspeed V, let us show that turbulence results when, for given Δt, a stable calculation results, but no J exists which yields a primary vortex.

Let us then set $V = -3000, \Delta t = 4(10)^{-7}, J = 75000$. The motion was simulated to $T = 0.60$. Typical results are shown at $T = 0.09, 0.12, 0.21,$ 0.27, 0.36, 0.48, 0.57, and 0.60 in Figs. 3.11–3.18, respectively. Figure 3.11 shows the very rapid development of a compression wave. Figures 3.12–3.14 show what appears to be a large primary vortex. However, Figs. 3.15–3.18 show the development of a strong vertical current in the left portion of the figure. Figure 3.18 shows that this current runs across the usual primary

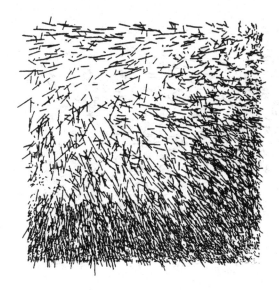

Fig. 3.11. $V = -3000$, $J = 75000$, $T = 0.09$.

Fig. 3.12. $V = -3000$, $J = 75000$, $T = 0.12$.

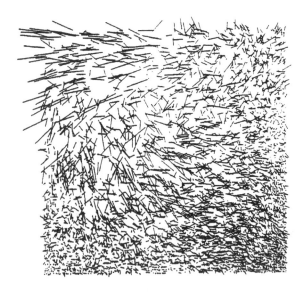

Fig. 3.13. $V = -3000$, $J = 75000$, $T = 0.21$.

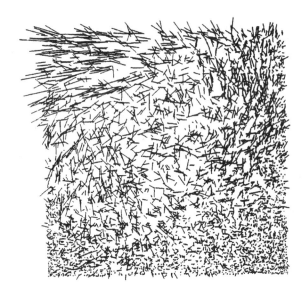

Fig. 3.14. $V = -3000$, $J = 75000$, $T = 0.27$.

Fig. 3.15. $V = -3000$, $J = 75000$, $T = 0.36$.

Fig. 3.16. $V = -3000$, $J = 75000$, $T = 0.48$.

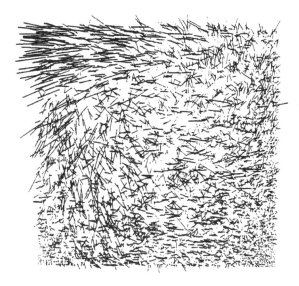

Fig. 3.17. $V = -3000,\ J = 75000,\ T = 0.57.$

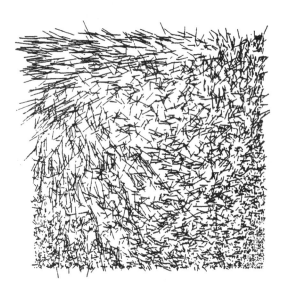

Fig. 3.18. $V = -3000,\ J = 75000,\ T = 0.6.$

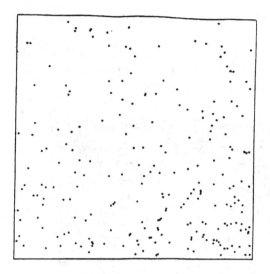

Fig. 3.19. 230 small vortices in Fig. 3.17.

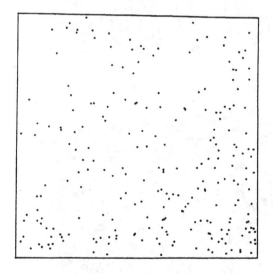

Fig. 3.20. 240 small vortices in Fig. 3.18.

vortex direction not only in the left portion of the figure but, to a lesser degree, also in the right portion. Thus, Criterion (1) is valid.

We again use the concept of a small vortex as in Section 2.6. For Fig. 3.19 shows that the flow in Fig. 3.17 has 230 small vortices at $T = 0.57$, while Fig. 3.20 shows that at $T = 0.60$, only 0.03 picoseconds later, the resulting flow has 240 small vortices which are completely different than those in Fig. 3.19.

Figures 3.21–3.23 show the same general flow as in Fig. 3.18, but using $J = 60000$, 45000, and 30000, respectively, so that the choice of J is immaterial to the development of turbulence.

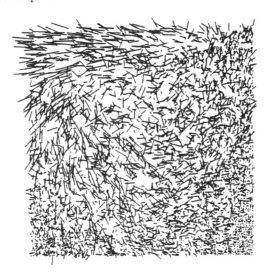

Fig. 3.21. $V = -3000$, $J = 60000$, $T = 0.60$.

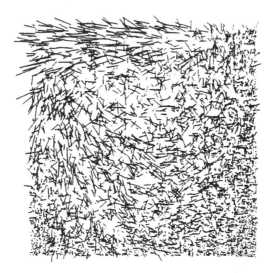

Fig. 3.22. $V = -3000$, $J = 45000$, $T = 0.60$.

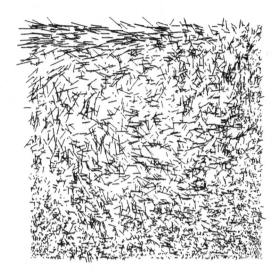

Fig. 3.23. $V = -3000$, $J = 30000$, $T = 0.60$.

3.6. The Fortran Program AIR.FOR

A typical computer program for the examples in this chapter is now given.

FORTRAN PROGRAM AIR.FOR

```
C  TOTAL NUMBER OF PARTICLES IS  4235.
c P AND Q ARE MOLECULAR PARAMETERS 7 AND 13.
      double precision XO(4235),YO(4235),VXO(4235),VYO(4235),
     1X(4235,3),Y(4235,3),VX(4235,2),VY(4235,2),
     1ACX(4235),ACY(4235),r2(4235),r4(4235),fx(4235),
     1fy(4235),f(4235)
     1,xke,R6(4235),R8(4235)
      OPEN (UNIT=21,file='air.dat',status='old')
      OPEN (UNIT=31,file='air.out',status='new')
      OPEN (UNIT=71,file='air.out1',status='new')
      OPEN (UNIT=72,file='air.out2',status='new')
      OPEN (UNIT=73,file='air.out3',status='new')
      OPEN (UNIT=74,file='air.out4',status='new')
      OPEN (UNIT=75,file='air.out5',status='new')
      K=1
      kk=1
      KPRINT=50000
```

```
        wallspeed= -3000.
        READ (21,10) (XO(I),YO(I),VXO(I),VYO(I),I=1,4235)
10      FORMAT (2f12.3,2f12.3)
        DO 30 I=1,4235
        X(I,1)=XO(I)
        Y(I,1)=YO(I)
        VX(I,1)=VXO(I)
        VY(I,1)=VYO(I)
30      CONTINUE
        GO TO 3456
C Cycle.
65      DO 70 I=1,4235
        X(I,1)=X(I,2)
        Y(I,1)=Y(I,2)
        VX(I,1)=VX(I,2)
        VY(I,1)=VY(I,2)
70      CONTINUE
        DO 701 I=1,4235
        ACX(I)=0.
        ACY(I)=0.
        F(I)=0.
701     CONTINUE
3456    DO 78 I=1,4234
        ACXI=ACX(I)
        ACYI=ACY(I)
        XI=X(I,1)
        YI=Y(I,1)
        IP1=I+1
        DO 77 J=IP1,4235
        R2(J)=(XI-X(J,1))**2+(YI-Y(J,1))**2
        R6(J)=R2(J)*R2(J)*R2(J)
        R8(J)=R6(J)*R2(J)
C in this program the local interaction distance is 2.5sigma.
c sigma = 3.6
C Force calculation.
        if (r2(j).gt.81.) go to 9000
        F(J)=149795.*(-1.+4478/(R6(J)))
1       /(R8(J))
        go to 9001
```

```
9000   f(j)=0.0
9001  ACX(J)=ACX(J)-F(J)*(XI-X(J,1))
      ACY(J)=ACY(J)-F(J)*(YI-Y(J,1))
      ACXI=ACXI+F(J)*(XI-X(J,1))
      ACYI=ACYI+F(J)*(YI-Y(J,1))
77    CONTINUE
      ACX(I)=ACXI
      ACY(I)=ACYI
78    CONTINUE
c Note, gravity is no consequence on the molecular level.
      DO 7123 I=1,4235
      VX(I,2)=VX(I,1)+0.0000004*ACX(I)
      VY(I,2)=VY(I,1)+0.0000004*ACY(I)
      X(I,2)=X(I,1)+0.0000004*VX(I,2)
      Y(I,2)=Y(I,1)+0.0000004*VY(I,2)
7123  CONTINUE
c at this point we insert wall reflection.
      do 995 i=1,4235
      if (x(i,2).lt.121.8) go to 992
      x(i,2)=243.6-x(i,2)
      vx(i,2)=-1.0*vx(i,2)
      vy(i,2)=0.0*vy(i,2)
992   if (x(i,2).gt.-121.8) go to 993
      x(i,2)=-243.6-x(i,2)
      vx(i,2)=-1.0*vx(i,2)
      vy(i,2)=0.0*vy(i,2)
993   if (y(i,2).gt.0.0) go to 994
      y(i,2)=-y(i,2)
      vx(i,2)=0.0*vx(i,2)
      vy(i,2)=-1.00*vy(i,2)
994   if (y(i,2).lt.242.604) go to 995
      y(i,2)=485.208-y(i,2)
      vx(i,2)=vx(i,2)+wallspeed
      vy(i,2)=-1.*vy(i,2)
995   continue
      K=K+1
      if (k.eq.2) go to 500
      go to 501
500   write (71,10) (x(i,2),y(i,2),vx(i,2),vy(i,2),i=1,4235)
```

```
501   If (k.eq.15000) go to 502
      go to 503
502   write (72,10) (x(i,2),y(i,2),vx(i,2),vy(i,2),i=1,4235)
503   If (k.eq.30000) go to 504
      go to 505
504   write (73,10) (x(i,2),y(i,2),vx(i,2),vy(i,2),i=1,4235)
505   If (k.eq.45000) go to 506
      go to 507
506   write (74,10) (x(i,2),y(i,2),vx(i,2),vy(i,2),i=1,4235)
507   if (k.eq.60000) go to 508
      go to 509
508   write (75,10) (x(i,2),y(i,2),vx(i,2),vy(i,2),i=1,4235)
509   if (k.eq.75000) go to 510
      go to 511
510   write (31,10) (x(i,2),y(i,2),vx(i,2),vy(i,2),i=1,4235)
511   continue
82    if (k.lt.75000) go to 65
      STOP
      END
```

Chapter 4

Molecular Cavity Flow of Water Vapor in Two Dimensions

4.1. Introduction

We turn attention now to water vapor and observe first (Horne (1981)) that "In the gaseous state, whether steam or vapor, water molecules are largely independent of one another, and, apart from collisions, interactions between them are slight. Gaseous water is largely monomeric (consists of single molecules.)". We assume then that the Rowlinson potential (Hirschfelder, Curtiss and Bird (1967), pp 1033–1034) can be approximated by

$$\phi(r_{ij}) = (1.9646)10^{-13} \left[\frac{2.725^{12}}{r_{ij}^{12}} - \frac{2.725^6}{r_{ij}^6} \right] \text{erg} \ \left(\frac{\text{grcm}^2}{\text{sec}^2} \right) \qquad (4.1)$$

in which r_{ij} is measured in angstroms (Å). The force $\vec{F}_{ij}$ exerted on P_i by P_j is then

$$\vec{F}_{ij} = (1.9646)10^{-5} \left[\frac{12(2.725^{12})}{r_{ij}^{13}} - \frac{6(2.725^6)}{r_{ij}^7} \right] \frac{\vec{r}_{ji}}{r_{ij}} \text{dynes} \ \left(\frac{\text{grcm}}{\text{sec}^2} \right).$$
$$\qquad (4.2)$$

Note that there is *no* known potential for *liquid* water. The primary reason being that close liquid water molecules exhibit hydrogen bonding.

4.2. Equations of Motion for Water Vapor Molecules

From the discussion in Section 4.1, it follows that the equation of motion for a single water vapor molecule P_i acted on by a single water vapor molecule

$P_j, i \neq j$, is

$$m_i \vec{a}_i = (1.9646)10^{-5} \left[\frac{12(2.725^{12})}{r_{ij}^{13}} - \frac{6(2.725^6)}{r_{ij}^7} \right] \frac{\vec{r}_{ji}}{r_{ij}}.$$

Since the mass of a water molecule is $(30.103)10^{-24}$ gr, the latter equation is equivalent to

$$\vec{a}_i = (160.33)10^{19} \left[\frac{818.90}{r_{ij}^{13}} - \frac{1}{r_{ij}^7} \right] \frac{\vec{r}_{ji}}{r_{ij}} \left(\frac{cm}{sec^2} \right).$$

Recasting the latter equation in $\text{Å}/(\text{ps}^2)$ yields

$$\vec{a}_i = (160330) \left[\frac{818.90}{r_{ij}^{13}} - \frac{1}{r_{ij}^7} \right] \frac{\vec{r}_{ji}}{r_{ij}} \left(\frac{\text{Å}}{\text{ps}^2} \right). \tag{4.3}$$

On the molecular level, however, the effective force on P_i is local, by which we will mean only those molecules within a distance D determined by $2.5\sigma = 2.5(2.725) = 6.8\,\text{Å}$ are to be considered. Thus, for $r_{ij} \geq D = 6.8\,\text{Å}$, we choose $\vec{F}_{ij} = \vec{0}$.

We now consider the cavity problem with $l = 183.6$. We generate 4235 points on a regular triangular grid with the recursion formulas:

$$x(1) = -91.8, \quad y(1) = 0$$
$$x(i) = x(i-1) + 3.06, \quad y(i) = 0, \quad i = 2, 61$$
$$x(62) = -90.27, \quad y(62) = 2.65$$
$$x(i) = x(i-1) + 3.06, \quad y(i) = 2.65, \quad i = 63, 121$$
$$x(i) = x(i-121), \quad y(i) = y(i-121) + 5.30, \quad i = 122, 4235.$$

At each point $(x(i), y(i))$ we set a water vapor molecule P_i. Thus the initial positions are known.

We determine the initial velocities to correspond to the temperature $15°C$ so that $v = 5.14\,\text{Å/ps}$. Each molecule is then assigned a speed of $5.14\,\text{Å/ps}$, at random, with its sign $(\pm)$ also determined at random. All computational considerations discussed in Chapter 2 are also applied here.

Observe also that for $V = -40\,\text{Å/ps}$, $B = 183.6\,\text{Å}$ and at $15°C, \nu = 113.12\,\text{Å}^2/\text{ps}$ (Streeter (1962)), then (2.11) yields $Re = 65$.

4.3. Examples of Primary Vortex Generation

For the parameter choices $V = -40, J = 15000, \Delta t = 0.0001$, Figs. 4.1–4.4 show the development of a primary vortex at the respective times

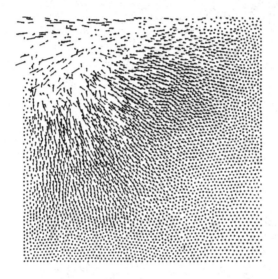

Fig. 4.1. $T = 2.5$.

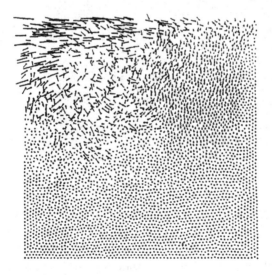

Fig. 4.2. $T = 5.0$.

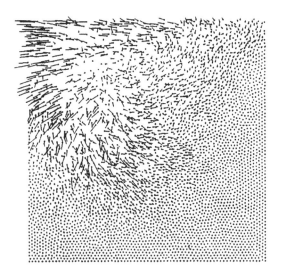

Fig. 4.3. $T = 7.5$.

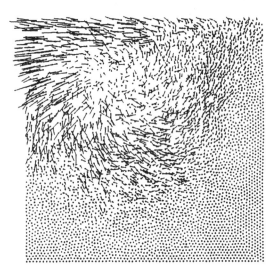

Fig. 4.4. $T = 10.0$.

$t = 2.5, 5.0, 7.5$, and 10.0. A very large region of compression develops and is seen readily in Fig. 4.1. The wall motion and the intermolecular repulsion which results from compression are the mechanisms by which cavity flow is generated. Other values of J which were studied were 25000, 20000, 10000 and 5000, but all yielded results entirely similar to those in Figs. 4.1–4.4.

Observe also that our figures agree qualitatively with the results of others when solving the Navier–Stokes equations numerically with $Re = 65$.

For the parameter choices $V = -100$ ($Re = 162$), $J = 5000$, $\Delta t = 0.0001$, Figs. 4.5, 4.6 show the development of a primary vortex at the times $t = 2.5, 5.0$. These figures reveal that at this wallspeed, the primary

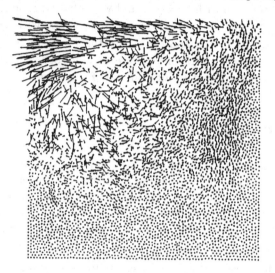

Fig. 4.5. $T = 2.5$.

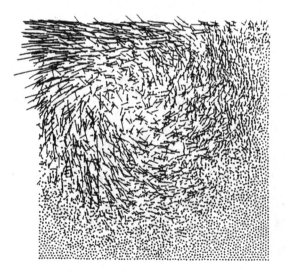

Fig. 4.6. $T = 5.0$.

vortex develops more quickly and is more extensive than in the previous example. The velocity fields in these figures were plotted with the same parameters as used in Figs. 4.1–4.4. Other values of J which were studied were 25000, 20000, 15000, and 10000, and all yielded results similar to those in Figs. 4.5 and 4.6.

Other examples were run with $V = -20$ and $V = -250$ with the expected results, that is, the time for the vortex to develop decreased with the absolute value of the magnitude of V and the size of the vortex increased with the absolute value of the magnitude of V. However, since our primary interest is in turbulence, we turn next to an example of turbulence.

4.4. Example of Turbulent Flow

Again, for a sufficiently large magnitude of the wallspeed V, let us show that for a given Δt, which yields a stable calculation, no J exists which yields a primary vortex. Hence, set $V = -3000(Re = 4869), \Delta t = 2(10)^{-6}, J = 50000$. Figures 4.7–4.11 show the resulting flows at $T = 4(10)^{-6}, 6(10)^{-6}, 8(10)^{-6}, 1(10)^{-5}, 1.2(10)^{-5}$, respectively. Figure 4.7 reveals extensive compression. However, Fig. 4.11 does not yield a primary vortex because of the presence of a strong crosscurrent across the primary

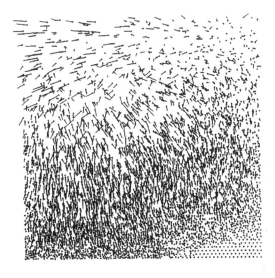

Fig. 4.7. $T = 4(10)^{-6}$.

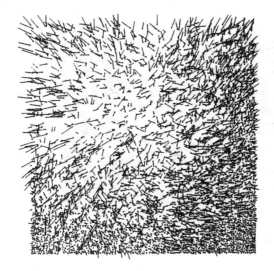

Fig. 4.8. $T = 6(10)^{-6}$.

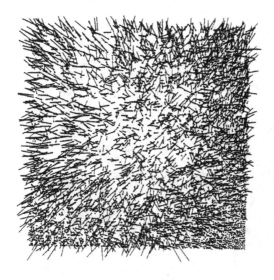

Fig. 4.9. $T = 8(10)^{-6}$.

vortex direction. Figure 4.11 shows full turbulence and the crosscurrent is easily seen in the figure all around the cavity.

Again, using the definition of a *small vortex*, Fig. 4.12 shows that the flow in Fig. 4.10 has 222 small vortices, while Fig. 4.13 shows that the flow in Fig. 4.11 has 220 small vortices, only $2(10)^{-6}$ picoseconds later.

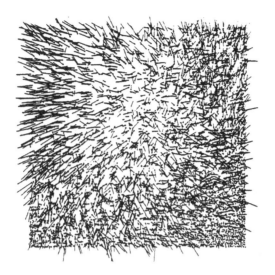

Fig. 4.10. $T = 1(10)^{-5}$.

Fig. 4.11. $T = 1.2(10)^{-5}$.

Moreover, the small vortices in Figure 4.11 are completely different than those in Fig. 4.10.

Other choices of J were 40000 and 30000. The results were entirely similar to those shown in Figs. 4.11, and are given in Figs. 4.14 and 4.15.

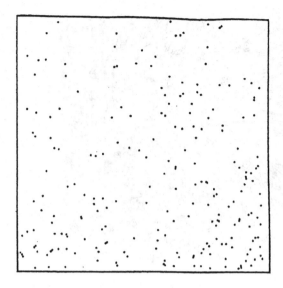

Fig. 4.12. 222 small vortices in Fig. 4.10.

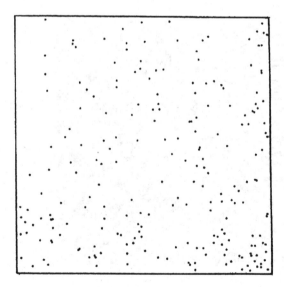

Fig. 4.13. 220 small vortices in Fig. 4.11.

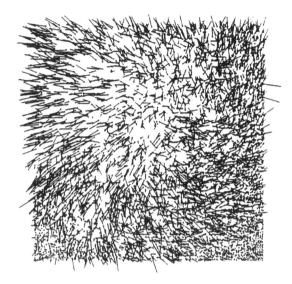

Fig. 4.14. $T = 1.2(10)^{-5}$, $J = 40000$.

Fig. 4.15. $T = 1.2(10)^{-5}$, $J = 30000$.

4.5. A Speculative Study of Liquid Water (Korzeniowski and Greenspan (1995))

Let us assume, for speculative purposes, that Eqs. (4.1)–(4.3) are approximate equations for the flow of liquid water and see what a probabilistic approach to the flow reveals. We will study only a small model of 446 molecules and use a very general qualitative characterization of turbulent flow, that is, a flow is turbulent if diffusive forces superimposed on laminar ones yield strong mixing (Schlichting (1960)). Also, for simplicity, we will consider only flow in a fixed direction, rather than cavity flow.

Consider 446 water molecules, at 15° C, as show in 9 rows in Fig. 4.16. The positions were generated by the recursion formulas:

$$x_1 = 0, \quad y_1 = 0$$
$$x_{i+1} = x_i + 3.05, \quad y_{i+1} = 0, \quad i = 1, 49$$
$$x_{51} = \frac{1}{2}(3.05), \quad y_{51} = 2.64$$
$$x_{i+1} = x_i + 3.05, \quad y_{i+1} = 2.64, \quad i = 51, 98$$
$$x_i = x_{i-99}, \quad y_i = y_i + 5.28, \quad i = 100, 446.$$

The initial velocities were chosen at random according to the Maxwell–Boltzmann law, in which a two dimensional root-mean-square velocity $v_{\mathrm{rms}} = (2kT/m)^{\frac{1}{2}}$ was used, with k, T, m being the Boltzmann constant, the absolute temperature, and the molecular mass, respectively. Finally a scalar stream velocity v, equal to the magnitude of the horizontal velocity in the x directon is superimposed to obtain the initial velocities $V_i^o = (V_{x,i} + v, V_{y,i})$. In our examples we choose $v = 10 \, \mathrm{m/sec}$.

Using the leap frog formulas with time step 0.0002, we discovered that the initial positions of the molecules played a major role in determining the onset of turbulence. Hence, because of its spatially isotropic nature, a circular Gaussian random vector $G_i = (G_{x,i}, G_{y,i})$ was chosen to yield $G_i^o = (x_i + G_{x,i}, y_i + G_{y,i})$ as the position perturbation of each water molecule from its original position (x_i, y_i). More precisely, our Gaussian random position

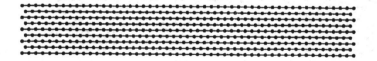

Fig. 4.16.   The initial configuration.

perturbation vector $G = (G_x, G_y)$ is chosen to be circular, zero mean with the probability density

$$g(x,y) = \frac{1}{2\pi} \times \frac{1}{s^2} e^{-(x^2 + y^2)/(2s^2)}, \quad s = \text{standard deviation.}$$

The standard deviation parameter s will be used as an indicator of the qualitative features of the flow.

Example 1. $s = 0.1\,\text{Å}$. The computer results are shown in Figs. 4.17–4.25 for 10000 time steps. In this case laminar flow is preserved.

Fig. 4.17.   400 steps.

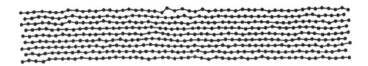

Fig. 4.18.   800 steps.

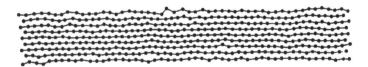

Fig. 4.19.   1200 steps.

Fig. 4.20.   1600 steps.

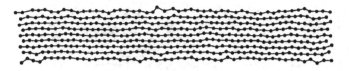

Fig. 4.21.   2000 steps.

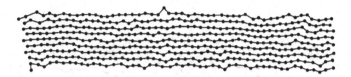

Fig. 4.22.   4000 steps.

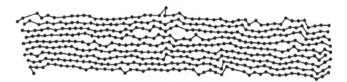

Fig. 4.23.   6000 steps.

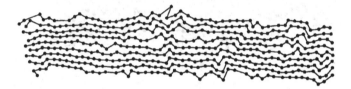

Fig. 4.24.   8000 steps.

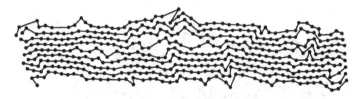

Fig. 4.25.   10000 steps.

Example 2. $s = 0.2$ Å. The results are shown in Figs. 4.26–4.30 which suggest a slow transition to turbulence.

Example 3. $s = 0.36$ Å. The results are shown in Figs. 4.31–4.33 and suggest a rapid transition to turbulent flow.

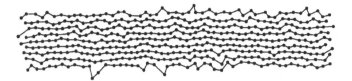

Fig. 4.26.   2500 steps.

Fig. 4.27.   5000 steps.

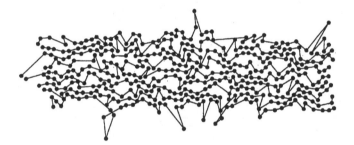

Fig. 4.28.   7500 steps.

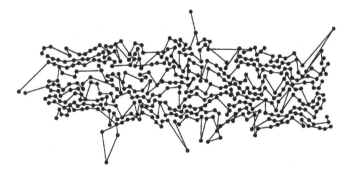

Fig. 4.29.   10000 steps.

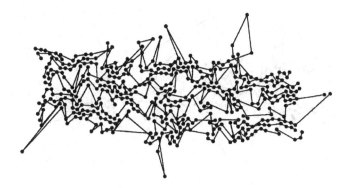

Fig. 4.30.   12500 steps.

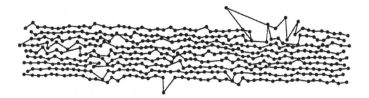

Fig. 4.31.   400 steps.

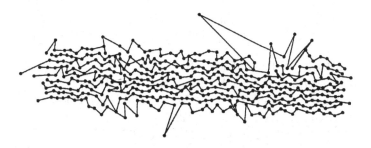

Fig. 4.32.   8000 steps.

Examples 1–3 lead us to conclude that for small position perturbations, a fraction of the intermolecular equilibrium distance, water retains the characteristics of laminar flow whereas, with the increase in the initial position perturbation, the flow becomes turbulent. Precisely, it was found that if the standard deviation s exceeds roughly 1/10 of the intermolecular equilibrium distance, then with an overwhelming probability of order 0.99, there

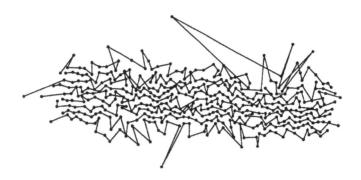

Fig. 4.33.   1200 steps.

is a transition to turbulent behavior. Moreover, the flow stayed laminar for perturbations of order less that 1/10 of the equilibrium separation.

In our discussion, we did not explore the origin of these perturbation, e.g. gravity, heating, friction, etc., and have confined our attenstion only to exploration of possible scenarios leading to turbulent flow.

4.6.   The Fortran Program CAV.FOR

A typical computer program for the examples in this chapter is now given.

FORTRAN PROGRAM CAV.FOR

```
c cav.FOR
C  TOTAL NUMBER OF PARTICLES IS  4235.
c P AND Q ARE MOLECULAR PARAMETERS 7 AND 13.
    double precision XO(4235),YO(4235),VXO(4235),VYO(4235),
    1X(4235,3),Y(4235,3),VX(4235,2),VY(4235,2),
    1ACX(4235),ACY(4235),r2(4235),r4(4235),fx(4235),
    1fy(4235),f(4235)
    1,xke,R6(4235),R8(4235)
    OPEN (UNIT=21,file='cav.dat',status='old')
    OPEN (UNIT=31,file='cav.out',status='new')
    OPEN (UNIT=71,file='cav.out1',status='new')
    OPEN (UNIT=72,file='cav.out2',status='new')
    OPEN (UNIT=73,file='cav.out3',status='new')
    OPEN (UNIT=74,file='cav.out4',status='new')
    OPEN (UNIT=75,file='cav.out5',status='new')
```

```
      K=1
      kk=1
      KPRINT=50000
      wallspeed= -100.
      READ (21,10) (XO(I),YO(I),VXO(I),VYO(I),I=1,4235)
10    FORMAT (2f12.3,2f12.3)
      DO 30 I=1,4235
      X(I,1)=XO(I)
      Y(I,1)=YO(I)
      VX(I,1)=VXO(I)
      VY(I,1)=VYO(I)
30    CONTINUE
      GO TO 3456
C Cycle.
65    DO 70 I=1,4235
      X(I,1)=X(I,2)
      Y(I,1)=Y(I,2)
      VX(I,1)=VX(I,2)
      VY(I,1)=VY(I,2)
70    CONTINUE
      DO 701 I=1,4235
      ACX(I)=0.
      ACY(I)=0.
      F(I)=0.
701   CONTINUE
3456  DO 78 I=1,4234
      ACXI=ACX(I)
      ACYI=ACY(I)
      XI=X(I,1)
      YI=Y(I,1)
      IP1=I+1
      DO 77 J=IP1,4235
      R2(J)=(XI-X(J,1))**2+(YI-Y(J,1))**2
      R6(J)=R2(J)*R2(J)*R2(J)
      R8(J)=R6(J)*R2(J)
C in this program the local interaction distance is 2.5sigma.
c sigma = 2.725
C Force calculation
      if (r2(j).gt.46.) go to 9000
```

```
      F(J)=160330.*(-1.+818.90/(R6(J)))
    1   /(R8(J))
      go to 9001
9000  f(j)=0.0
9001  ACX(J)=ACX(J)-F(J)*(XI-X(J,1))
      ACY(J)=ACY(J)-F(J)*(YI-Y(J,1))
      ACXI=ACXI+F(J)*(XI-X(J,1))
      ACYI=ACYI+F(J)*(YI-Y(J,1))
77    CONTINUE
      ACX(I)=ACXI
      ACY(I)=ACYI
78    CONTINUE
c Note, gravity is no consequence on the molecular level.
      DO 7123 I=1,4235
      VX(I,2)=VX(I,1)+0.0001*ACX(I)
      VY(I,2)=VY(I,1)+0.0001*ACY(I)
      X(I,2)=X(I,1)+0.0001*VX(I,2)
      Y(I,2)=Y(I,1)+0.0001*VY(I,2)
7123  CONTINUE
c at this point we insert wall reflection.
      do 995 i=1,4235
      if (x(i,2).lt.91.8) go to 992
      x(i,2)=183.6-x(i,2)
      vx(i,2)=-1.0*vx(i,2)
      vy(i,2)=0.0*vy(i,2)
992   if (x(i,2).gt.-91.8) go to 993
      x(i,2)=-183.6-x(i,2)
      vx(i,2)=-1.0*vx(i,2)
      vy(i,2)=0.0*vy(i,2)
993   if (y(i,2).gt.0.0) go to 994
      y(i,2)=-y(i,2)
      vx(i,2)=0.0*vx(i,2)
      vy(i,2)=-1.00*vy(i,2)
994   if (y(i,2).lt.183.6) go to 995
      y(i,2)=367.2-y(i,2)
      vx(i,2)=vx(i,2)+wallspeed
      vy(i,2)=-1.*vy(i,2)
995   continue
      K=K+1
```

```
      if (k.eq.2) go to 500
      go to 501
500   write (71,10) (x(i,2),y(i,2),vx(i,2),vy(i,2),i=1,4235)
501   If (k.eq.5000) go to 502
      go to 503
502   write (72,10) (x(i,2),y(i,2),vx(i,2),vy(i,2),i=1,4235)
503   If (k.eq.10000) go to 504
      go to 505
504   write (73,10) (x(i,2),y(i,2),vx(i,2),vy(i,2),i=1,4235)
505   If (k.eq.15000) go to 506
      go to 507
506   write (74,10) (x(i,2),y(i,2),vx(i,2),vy(i,2),i=1,4235)
507   if (k.eq.20000) go to 508
      go to 509
508   write (75,10) (x(i,2),y(i,2),vx(i,2),vy(i,2),i=1,4235)
509   if (k.eq.25000) go to 510
      go to 511
510   write (31,10) (x(i,2),y(i,2),vx(i,2),vy(i,2),i=1,4235)
511     continue
82    if (k.lt.25000) go to 65
      STOP
      END
```

Chapter 5

Molecular Cavity Flow of Water Vapor in Three Dimensions

5.1. Introduction

To simulate the flow of water vapor in three dimensions, vector Eq. (4.3) remains valid and for convenience is recalled now as

$$\frac{d^2\vec{r}_i}{dt^2} = (160330) \sum_{\substack{j \\ j \neq i}} \left[\frac{818.90}{r_{ij}^{13}} - \frac{1}{r_{ij}^7} \right] \frac{\vec{r}_{ji}}{r_{ij}}; \quad i = 1, 2, 3, \ldots, N; \quad r_{ij} < D.$$

(5.1)

We will have to formulate a three dimensional cavity problem and will find that the numerical solution of the dynamical equations can be exceptionally arduous. The computations described in this chapter were executed on a Digital Alpha 533, but several "tricks" had to be introduced in order to obtain results in a reasonable time period.

Recall also that associated with (5.1) are the equilibrium distance 3.06 Å and the interaction distance D. The local interaction distance D is now taken to be the solution of the equation

$$\frac{dF_{ij}}{dr_{ij}} = 0,$$

the solution of which is $r_{i,j} = 3.39$ Å. This choice not only saves computer time, but stresses the fact that, in turbulent flow, the major force is repulsion, with the effect of attraction being minimalized. As an additional time saver, the force is not calculated anew at each time step, but at every other time step. This is reasonable because the time step in the turbulent calculation is exceptionally small.

It is most important to note that forces exist in three dimensional cavity flow which are not present in two dimensions (Roache (1972)).

5.2.   Molecular Arrangement and the Cavity Problem

Using the equilibrium value 3.06 Å, we first construct a cubic grid of points as follows. Consider the three dimensional cube with

$$-52.02 \leq x \leq 52.02, \quad 0 \leq y \leq 104.04, \quad 0 \leq z \leq 104.04,$$

as shown in Fig. 5.1. The edge of this cube has length 104.04 Å and it is symmetrical about the YZ plane, that is, about the plane $x = 0$. In this cube a three dimensional cubic grid is constructed with $\Delta x = \Delta y = \Delta z = 3.06$ Å as follows:

$$x(1) = -52.02, \quad y(1) = 0.0, \quad z(1) = 0.0$$
$$x(i+1) = x(i) + 3.06, \quad y(i) = 0.0, \quad z(i) = 0.0, \quad i = 1, 34$$
$$x(36) = 0.0, \quad y(36) = 3.06, \quad z(36) = 0.0$$
$$x(i+1) = x(i) + 3.06, \quad y(i+1) = 3.06, \quad z(i+1) = 0.0, \quad i = 36, 69$$
$$x(i) = x(i - 70), \quad y(i) = y(i - 70) + 6.12, \quad z(i) = 0.0, \quad i = 71, 1225$$
$$x(i) = x(i - 1225), \quad y(i) = y(i - 1225), \quad z(i) = 3.06, \quad i = 1226, 2450$$
$$x(i) = x(i - 2450), \quad y(i) = y(i - 2450),$$
$$z(i) = z(i - 2450) + 6.12, \quad i = 2451, 42872.$$

There result 42875 interior and boundary grid points of which 1225 lie in the YZ plane. Renumber the points in the YZ plane to be

$$x(i), y(i), z(i); \quad i = 20826, \ 22050.$$

Renumber the remaining points so that

$$x(i) = -x(42876 - i), \quad y(i) = y(42876 - i), \quad z(i) = z(42876 - i),$$
$$i = 22051, \ 42875.$$

At grid point numbered i, set a molecule P_i; $i = 1, 42875$. To complete the initial data for the molecules, note that at 150°C, the speed in Å/ps for a water molecule in three dimensions, from (1.5), is 7.6 Å/ps. This speed and its sign are assigned to each molecule in the X, Y, or Z direction, at random, but with the reasonable restrictions (Freitas, Street, Findikakis, and Koseff (1985)) that if P_i and P_j are not in the YZ plane, but are

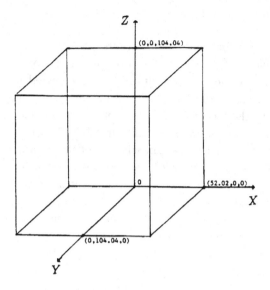

Fig. 5.1.   The cubic basin.

symmetrical relative to the YZ plane, then

$$v_x(P_i) = -v_x(P_j), \quad v_y(P_i) = v_y(P_j), \quad v_z(P_i) = v_z(P_j),$$

while if P_i is in the YZ plane then $v_x(P_i) = 0$. These assumptions are allowable because the cavity problem to be described yields, experimentally, a flow which is symmetrical about the YZ plane. In this fashion, through the use of symmetry , the dynamics reduces to a 22050-body problem, and hence is an additional time saver.

For the 42875 water molecules with dynamical Eq. (5.1) and the initial data described above, we now describe the three dimensional cavity, or, basin, problem to be studied. The top face of the cube is the only face allowed to move. It moves in the Y direction and is allowed an extended length so that the fluid is always enclosed by six faces. The constant speed V of the top face is called the wallspeed. Our problems are to describe the gross fluid motion within the cube for various values of V, and, of course, to produce turbulence.

5.3. Computational Considerations

In applying the leap frog formulas, we must first describe a protocol to apply when a molecule has crossed a face into the exterior of the cavity. For each of the lower five faces, the position will be reflected back symmetrically relative to that face into the interior of the cavity, the velocity components tangent to the face will be set to zero, and the velocity component perpendicular to the face will be multiplied by -1. If the molecule crosses the moving face, then its position will be reflected back symmetrically, its x component of velocity will be set to zero, its z component of velocity will be multiplied by -1, and its y component of velocity will be increased by the wallspeed V.

Again, since the molecular motion is Brownian, we introduce average velocities $\vec{v}_{i,k,J}$ as follows. For J a positive integer, let molecule P_i be at $(x(i,k), y(i,k), z(i,k))$ at $t_k = k\Delta t$ and at $(x(i,k-J), y(i,k-J), z(i,k-J))$ at $t_{k-J} = (k-J)\Delta t$. Then, the average velocity $\vec{v}_{i,k,J}$ of P_i at t_k is defined by

$$\vec{v}_{i,k,J} = \left(\frac{x(i,k) - x(i,k-J)}{J\Delta t}, \frac{y(i,k) - y(i,k-J)}{J\Delta t}, \frac{z(i,k) - z(i,k-J)}{J\Delta t} \right).$$

$$(5.2)$$

5.4. Examples

Consider first $V = 25, \Delta t = 0.00008$, and $J = 25200$. Figures 5.2–5.6 show vortex development for $x = 0$, that is, in the YZ plane at the respective times $t = 2.90, 3.35, 4.02, 7.04, 10.07$. Figure 5.2 reveals the development of a compression wave moving down and to the left. This results in compressioncompression and molecular repulsion upwards, as is apparent in the upper left sections of Figs. 5.3 and 5.4. Continuation of this motion results in the vortex flows shown in Figs. 5.5 and 5.6. For graphical clarity, the velocities are reduced by the factor $vm = 0.51$. Calculations using $J = 19600, 22400, 28000, 30800$ yielded results entirely similar to those displayed in Figs. 5.2–5.6.

Motion of molecules not in the YZ plane was often difficult to display. Those to be described next were for the time interval $0 \le t \le 8.912 \, \text{ps}$.

The molecule P_{17000}, labeled A in Fig. 5.7, is initially at $(-3.06, 58.14, 87.68)$. After extensive local interaction, the molecule moves to $(-2.21, 33.61, 81.35)$, which is labeled B in Fig. 5.7. It then moves rapidly in the Z direction to $(-2.42, 24.23, 100.32)$, labeled C in the figure. Both X and Y

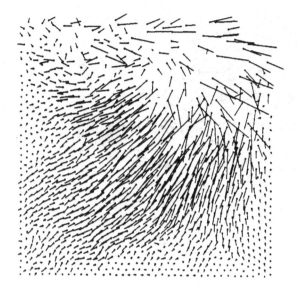

Fig. 5.2. $T = 2.90$.

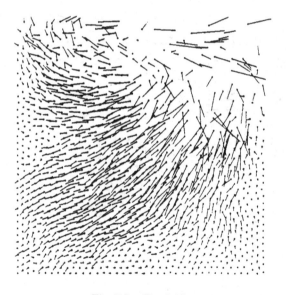

Fig. 5.3. $T = 3.35$.

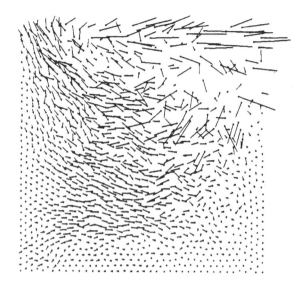

Fig. 5.4. $T = 4.02$.

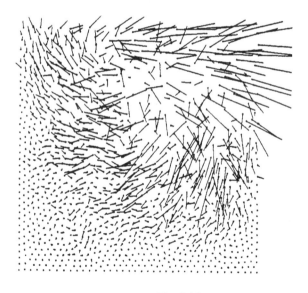

Fig. 5.5. $T = 7.04$.

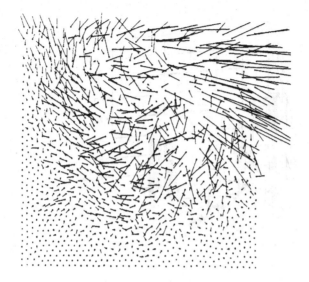

Fig. 5.6. $T = 10.07$.

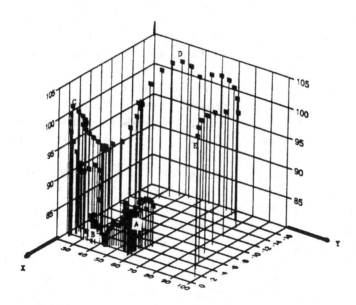

Fig. 5.7. P_{20500}.

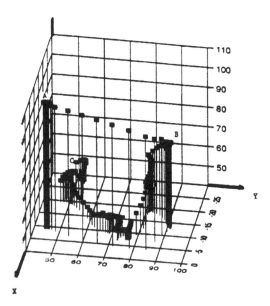

Fig. 5.8. P_{20500}.

values change rapidly next as it moves to $(-17.24, 38.61, 102.37)$, labeled D in the figure. It relocates finally with rapid motion in the X and Y directions to $(-1.37, 98.10, 103.48)$, shown as E in the figure.

The molecule P_{20500}, labeled A in Fig. 5.8 is initially at $(-9.18, 45.90, 104.04)$. It moves relatively rapidly, primarily in the Y direction, to $(-14.07, 89.81, 79.32)$, labeled B in the figure. It then descends in a somewhat spiral fashion to its terminal point $(-25.74, 50.31, 52.15)$, shown as C in the figure.

The molecule P_{22500}, labeled A in Fig. 5.9, is at $(24.48, 24.48, 104.04)$ initially. Its initial motion is rapid in the Y direction to $(19.13, 93.83, 102.7)$, shown as B in the figure. It then descends to $(21.29, 78.43, 57.96)$, shown as C in the figure, and finally it sidles its way down to $(2.74, 72.29, 46.76)$.

The molecule P_{23500} begins at $(15.30, 58.14, 97.92)$, shown as A in Fig. 5.10. It moves rapidly in the Y direction to $(35.16, 99.32, 99.51)$, shown as B in the figure. From there it sidles its way downward in an oscillatory fashion to its final point $(50.96, 95.27, 66.07)$, shown as C in the figure.

In each of the above examples, extensive motion results because the initial point is relatively close to the moving face. Molecular motion in the bottom of the cavity was far less extensive. Motion in the corners usually appeared to consist of random local oscillations.

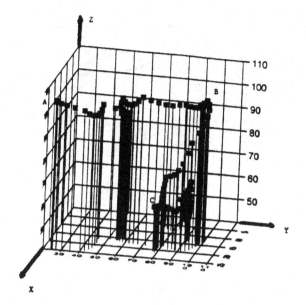

Fig. 5.9. P_{22500}.

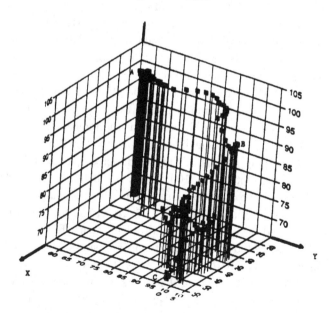

Fig. 5.10. P_{23500}.

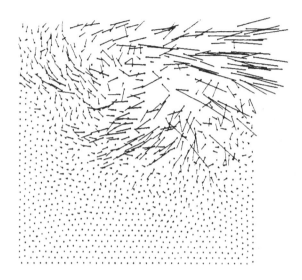

Fig. 5.11.   Two dimensional simulation at $T = 7.04$.

Figure 5.11 shows a direct, two dimensional simulation of the flow in the YZ plane for wallspeed 25, $J = 25200$, and $\Delta t = 0.00008$ at $T = 7.04$. The initial data are exactly those for the molecules in the YZ plane in the three dimensional example described above. The flow is analyzed using the very same parameters as those used to display Fig. 5.5. Comparison of the results in Fig. 5.5 with those in Fig. 5.11 show that there is a distinct difference in the two dimensional and three dimensional simulations. The average speed of the molecules in Fig. 5.11 is $4.3574\,\text{Å/ps}$, while those in Fig. 5.5 is $8.2368\,\text{Å/ps}$. In addition, the molecular distributions, though similar, are different. The differences are the result of the three dimensional forces which play no role in the two dimensional simulation.

For our next example, set $V = 100$, $\Delta t = 0.00004$, $J = 14000$. Figures 5.12–5.16 show the vortex development in the YZ plane at the respective times $t = 2.90, 3.35, 4.02, 7.04, 10.07$. As in the first example, the velocities were reduced in the figures for graphical clarity. Results using $J = 11200, 16800, 19600, 22400$ yielded results entirely similar to those for $J = 14000$. The mean speed at 10.07 in Fig. 5.16 is $10.56\,\text{Å/ps}$. This result, when compared with the mean speed of $9.46\,\text{Å/ps}$, for wallspeed $25\,\text{Å/ps}$ at 10.07, indicates that an increase in wallspeed by a factor of four need not result in a proportionate increase in average fluid speed throughout the cavity. This is probably due to the strong damping at the walls and due to the fact that much of the lower half of the cavity still has not been affected

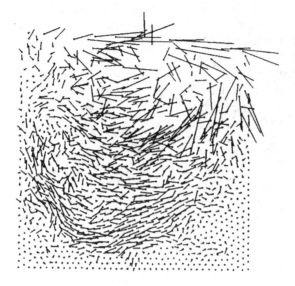

Fig. 5.12. $T = 2.90$.

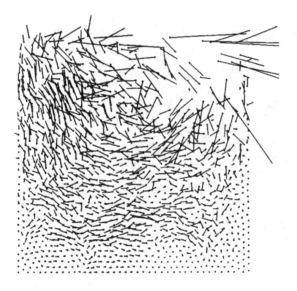

Fig. 5.13. $T = 3.35$.

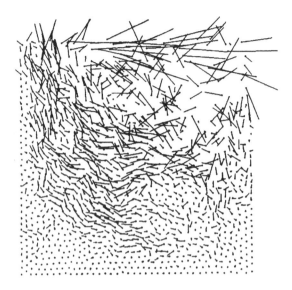

Fig. 5.14. $T = 4.02$.

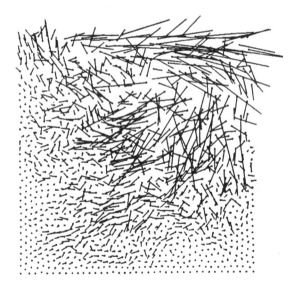

Fig. 5.15. $T = 7.04$.

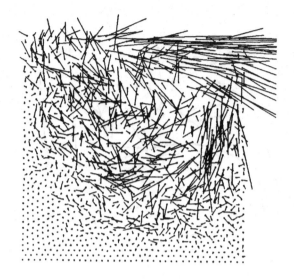

Fig. 5.16. $T = 10.07$.

significantly by the motion of the upper wall. Indeed, in Fig. 5.6 the lower half of the cavity has 747 molecules with a mean speed of 4.63 Å/ps, while in Fig. 5.16 the lower half of the cavity has 741 molecules with a mean speed of 4.69 Å/ps.

Though individual molecular trajectories showed more motion than that displayed in Fig. 5.7, no results of particular interest, like the appearance of Taylor–Gortler vortices, were observed. This is probably due to the relatively low wallspeed. For variety, then, let us display the secondary vortices in Fig. 5.16. Since secondary vortices have rotational speeds much smaller than for primary vortices, it is necessary to use relatively large values of J to display them. Thus, for $J = 56000$, Fig. 5.17 shows the secondary vortex in the lower right hand corner, while for $J = 50000$, Fig. 5.18 shows a smaller vortex in the lower left corner. These results are consistent with experimental results (Pan and Acrivos (1967)). We were able to find only one secondary vortex for $V = 25$, and this was in the lower right corner.

5.5.   Turbulent Flow

In turning to turbulent flow we will concentrate on $V = 6500$. However, it soon became apparent that for $V = 6500$, the problem of roundoff error accumulation required the time step $\Delta t = 2(10)^{-7}$. The projected time in

Fig. 5.17.   Lower right secondary vortex.

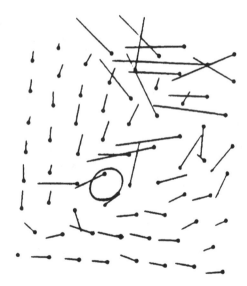

Fig. 5.18.   Lower left secondary vortex.

order to reach meaningful results on our Digital Alpha 533 was unreasonable. So, the decision was made to decrease the number of molecules. Thus a cubic grid of 29791 points was generated as follows:

$$x(1) = -45.9, \quad y(1) = 0.0, \quad z(1) = 0.0$$
$$x(i+1) = x(i) + 3.06, \quad y(i) = 0.0, \quad z(i) = 0.0, \qquad i = 1, 30$$
$$x(32) = 0.0, \quad y(32) = 3.06, \quad z(32) = 0.0$$
$$x(i+1) = x(i) + 3.06, \quad y(i+1) = 3.06, \quad z(i+1) = 0.0, \qquad i = 32, 61$$
$$x(i) = x(i-62), \quad y(i) = y(i-62) + 6.12, \quad z(i) = 0.0, \qquad i = 63, 961$$
$$x(i) = x(i-961), \quad y(i) = y(i-961), \quad z(i) = 3.06, \qquad i = 962, 1922$$
$$x(i) = x(i-1922), \quad y(i) = y(i-1922),$$
$$z(i) = z(i-1922) + 6.12, \qquad i = 1923, 29791.$$

There result 29791 interior and boundary grid points of which 961 lie in the YZ plane. Renumber the points in the YZ plane to be

$$x(i), y(i), z(i); \quad i = 14416, 15376.$$

Renumber the remaining points so that

$$x(i) = -x(29792 - i), \quad y(i) = y(29792 - i), \quad z(i) = z(29792 - i),$$
$$i = 15377, 29791.$$

At grid point numbered i, set a molecule P_i; $i = 1, 29791$. To complete the initial data for the molecules, recall that at 150°C, the speed in Å/ps for a water molecule in three dimensions, from (1.5), is 7.6 Å/ps. This speed and its sign are assigned to each molecule in the X, Y, or Z direction, at random, but with the restrictions that if P_i and P_j are not in the YZ plane, but are symmetrical relative to the YZ plane, then

$$v_x(P_i) = -v_x(P_j), \quad v_y(P_i) = v_y(P_j), \quad v_z(P_i) = v_z(P_j),$$

while if P_i is in the YZ plane then $v_x(P_i) = 0$. The resulting symmetry reduces the problem to a 15376-body problem.

Figures 5.19–5.21 show the very rapid development of turbulence in the YZ plane at the times $t = 0.0561, 0.1100, 0.1650$, with $J = 77000$. Figure 5.21 shows very clearly the large crosscurrent over the center bottom, center right, center left, and top left sections of the usual primary vortex direction. Entirely similar results followed for $J = 44000, 55000, 66000, 88000$.

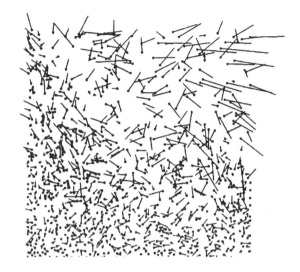

Fig. 5.19. $T = 0.0561$.

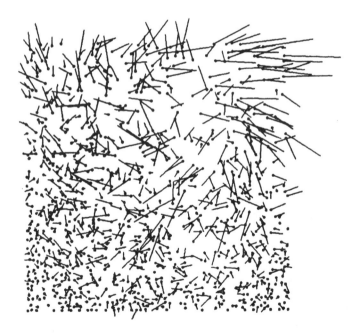

Fig. 5.20. $T = 0.1100$.

Fig. 5.21. $T = 0.1650$.

Figures 5.22–5.25 show typical molecular motions through the trajectories of $P_{4000}, P_{13500}, P_{19000}, P_{29500}$ during the time interval $0.0627 \leq t \leq 0.1914$. The initial point of P_{4000} in Fig. 5.22 is $(-26.37,\ 64.84,\ 1.60)$ and its terminal point is $(-3.08,\ 90.23,\ 24.51)$. The motion is extensive for such a short period of time and is characterized by multiple positions of unstable equilibrium. The motion of P_{13500} in Fig. 5.23 begins at $(-3.06, 0.00, 88.92)$ and terminates at $(-2.65, 61.43, 40.17)$. The motion of P_{19000} in Fig. 5.24 shows the same complexities as those in Fig. 5.22 and 5.23. Figure 5.25 shows that even the molecules at the bottom of the cavity have extensive motion. The motion of P_{29500} is entirely in the plane $Z = 0$. It begins at $(71.44, 13.33, 0.00)$ and terminates at $(26.65, 35.04, 0.00)$

Figures 5.26–5.28 show the results at the times $t = 0.06, 0.12$, and 0.18 with $J = 60000$ for the wallspeed 3500. Figure 5.28 clearly shows the development of turbulence in the center bottom portion of the YZ plane at $t = 0.18$. A strong crosscurrent has developed at this time. Entirely similar results followed for $J = 40000, 50000, 70000$.

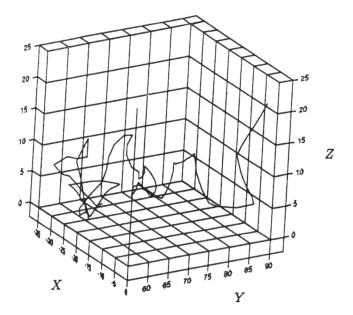

Fig. 5.22. P_{4000}.

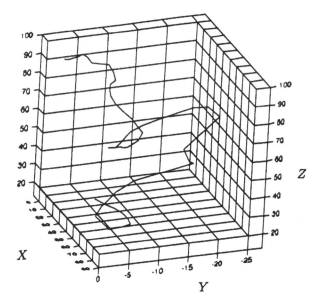

Fig. 5.23. P_{13500}.

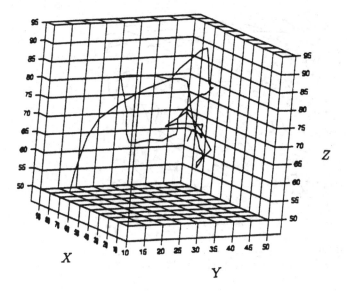

Fig. 5.24. P_{19000}.

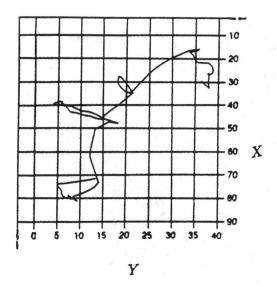

Fig. 5.25. P_{29500}.

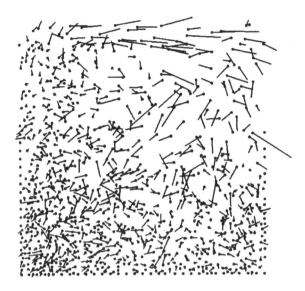

Fig. 5.26. $T = 0.06$.

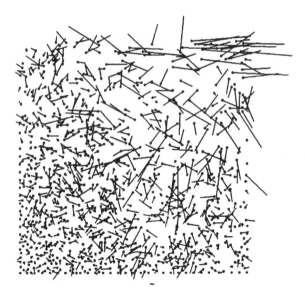

Fig. 5.27. $T = 0.12$.

Fig. 5.28. $T = 0.180$.

5.6. The Fortran program CAV3D.for.

In this section we give a typical Fortran program for the examples discussed in this chapter.

FORTRAN 90 PROGRAM CAV3D.FOR

```
C Total number of particles is 29791.
C Molecular parameters p and q are 7 and 13
C This is a vectorized program.  Subscripts 1,2,3 are X, Y, Z.
      PARAMETER (NP=29791,NH=15376,ND=3)
      IMPLICIT REAL*8(A-H,Q-Z)
      DIMENSION X(NP,ND),VX(NP,ND),ACX(NP,ND)
      OPEN (UNIT=21,FILE='C31A.DAT',STATUS='OLD')
      OPEN (UNIT=31,FILE='C31A.OUT')
      OPEN (UNIT=74,FILE='C31A.XZ31')
      OPEN (UNIT=75,FILE='C31A.TI31')
      OPEN (UNIT=76,FILE='C31A.TR31')
      K=0
      XSEC=SECNDS(0.0)
```

```
         KK=1
         KPRINT=2
         WALLSPEED=6500.
         READ (21,'(6F11.3)')(X(I,:),VX(I,:),I=1,NP)
65       CONTINUE
         IF (MOD(K,KPRINT).GT.0) GO TO 7777
         ACX=0.0
         DO I=1,NH
              DO J=I+1,NP
                   R2=SUM((X(I,:)-X(J,:))**2)
C  In this program the local interaction distance is 3.39
                   IF (R2.GT.11.50) CYCLE
                   IF (R2.EQ.0.0) GO TO 1998
                   R6=R2*R2*R2
                   R8=R6*R2
                   F=(160330.*(-1.+818.96/R6))/R8
                   ACX(I,:)=ACX(I,:)+F*(X(I,:)-X(J,:))
                   IF (J<=NH)ACX(J,:)=ACX(J,:)-F*(X(I,:)-X(J,:))
              END DO
         END DO
C Note that gravity is of no consequence on the molecular level.
7777     CONTINUE
         VX(1:NH,:)=VX(1:NH,:)+0.0000002*ACX(1:NH,:)
         X(1:NH,:)=X(1:NH,:)+0.0000002*VX(1:NH,:)
C  At this point we insert wall reflection.
C  We first reflect x values.
         WHERE(ABS(X(1:NH,1))>=45.9)
              X(1:NH,1)=MAX(-91.8-X(1:NH,1),
                     MIN(91.8-X(1:NH,1),X(1:NH,1)))
              VX(1:NH,1)=-VX(1:NH,1)
              VX(1:NH,2)=0.0
              VX(1:NH,3)=0.0
         END WHERE
C We next reflect y values.
         WHERE(ABS(X(1:NH,2)-45.9)>=45.9)
              X(1:NH,2)=MAX(-X(1:NH,2),
                     MIN(183.6-X(1:NH,2),X(1:NH,2)))
              VX(1:NH,2)=-VX(1:NH,2)
              VX(1:NH,1)=0.0
```

```
          VX(1:NH,3)=0.0
     END WHERE
C Next z values are reflected.
     WHERE(X(1:NH,3)<=0.0)
          X(1:NH,3)=-X(1:NH,3)
          VX(1:NH,3)=-VX(1:NH,3)
          VX(1:NH,1)=0.0
          VX(1:NH,2)=0.0
     END WHERE
     WHERE(X(1:NH,3)>=91.8)
          X(1:NH,3)=183.6-X(1:NH,3)
          VX(1:NH,1)=0.0
          VX(1:NH,2)=WALLSPEED
          VX(1:NH,3)=-VX(1:NH,3)
     END WHERE
     X (NH+1:NP,1    )=-X (14415: 1:-1,1 )
     VX(NH+1:NP,1    )=-VX(14415:1:-1,1 )
     X (NH+1:NP,2:3  )=X (14415: 1:-1,2:3)
     VX(NH+1:NP,2:3 )=VX(14415:1:-1,2:3)
          K=K+1
     IF (K.LT.5500) GO TO 65
     XXSEC=SECNDS(XSEC)
     WRITE (75,'(F20.1)')XXSEC
     WRITE (31,'(6F11.3)')(X(I,:),VX(I,:),I=1,NP)
     WRITE (74,'(4F11.3)')(X(I,2:3),VX(I,2:3),I=14416,15376)
     DO 1776 I=1,80
     X(I,:)=X(500*I,:)
     VX(I,:)=VX(500*I,:)
     WRITE (76,'(I12,6F11.3)')500*I,X(I,:),VX(I,:)
1776 CONTINUE
     STOP
1998 WRITE (75,'(3I7,6F11.3)')K,I,J,X(I,:),X(J,:)
     WRITE (31,'(6F11.3)') (X(I,:),VX(I,:),I=1,NP)
     END
```

Chapter 6

Particle Models of Flow in Two Dimensions

6.1. Introduction

In order to simulate fluid motion in the large, in this chapter we will introduce *particle modelling*. The essential idea derives from the popular *lumped mass* methodology in engineering. Hence, we will accumulate molecules into large entities called particles using mass and energy conservation. The particles will be studied dynamically by means of molecular type formulas, in which it will be essential to include gravity.

6.2. Particle Arrangement and Equations

Consider first a 30 cm by 240 cm rectangle and on it construct a regular triangular grid with 8479 points, shown in Fig. 6.1, by the recursion formulas

$$x(1) = -15.0, \quad y(1) = 0.0$$
$$x(i) = x(i-1) + 1.0, \quad y(i) = 0.0, \quad i = 2, 31$$
$$x(32) = -14.5, \quad y(32) = 0.866$$
$$x(i) = x(i-1) + 1.0, \quad y(i) = 0.866, \quad i = 33, 61$$
$$x(i) = x(i-61), \quad y(i) = y(i-61) + 1.732, \quad i = 62, 8479.$$

The side of each triangle in the grid has length 1 cm. At each grid point (x_i, y_i) set a particle P_i, that is an aggregate of molecules. Thus, the distance between any two immediate neighbors is unity.

Each particle is assigned a speed of 0.0001, with its sign and direction determined completely at random.

Fig. 6.1.   8479 particles.

Now we wish the particles to behave somewhat like molecules in the sense that close particles will repel while more distant ones will attract. However, since particles will be relatively massive, we do not wish them to exhibit the volatile motion common to molecules. Hence, the force between two distinct particles P_i and P_j which are R_{ij} cm apart will be taken to have magnitude F given by

$$F(R_{ij}) = -\frac{A}{R_{ij}^3} + \frac{B}{R_{ij}^5} \text{ (dynes)}. \tag{6.1}$$

Thus,

$$\phi(R_{ij}) = -\frac{A}{2R_{ij}^2} + \frac{B}{4R_{ij}^4} \text{ (ergs)}. \tag{6.2}$$

Our first problem is to determine A and B. Assume that $F(1) = 0$, so that from (6.1),

$$-A + B = 0. \tag{6.3}$$

In order to determine a second equation, some relevant observations must be made first.

Let us direct attention first to water vapor molecules. Note that the number N of molecules which can be arranged in the rectangle using the regular triangular grid approach is

$$N = \frac{30}{(3.06)10^{-8}} \cdot \frac{240}{(2.65)10^{-8}} = (8.87)10^{18}. \tag{6.4}$$

Also, note that since the mass of a water vapor molecule is $(30.103)10^{-24}$ gr, the total mass M of the water molecules inside the 30 cm by 240 cm rectangle in Fig. 6.1 is

$$M = (2.67)10^{-4} \text{ gr}. \tag{6.5}$$

Distributing this mass over the 8479 particles for conservation of total mass yields an individual particle mass m of

$$m = (3.15)10^{-8} \text{ gr}. \tag{6.6}$$

Assign each molecule a speed of 6.23 Å/ps with its direction and sign determined at random.

From (4.1) and (6.4), the total potential energy E_M of the molecular configuration is, approximately,

$$E_M = 3 \sum_{i=1}^{(8.88)10^{18}} \left\{ (1.9646)10^{-13} \left[\left(\frac{2.725}{3.06} \right)^{12} - \left(\frac{2.725}{3.06} \right)^6 \right] \right\}$$

$$= -(1.3)10^6 \, \text{erg}. \tag{6.7}$$

On the other hand, the total potential energy E_p of the particle configuration is, from (6.2), approximately,

$$E_p = 3 \sum_{i=1}^{8479} \left(-\frac{1}{2}A + \frac{1}{4}B \right) = 25437 \left(-\frac{1}{2}A + \frac{1}{4}B \right). \tag{6.8}$$

Now, to determine a second equation for A and B, we need only note that the kinetic energies of both the molecular and particle configurations are relatively small, so that we set $E_M = E_p$. Thus, from (6.7) and (6.8), the second equation for A and B is

$$25437 \left(-\frac{1}{2}A + \frac{1}{4}B \right) = -(1.3)10^6. \tag{6.9}$$

The solution of (6.3) and (6.9) is $A = B = 205$. Thus, (6.1) takes the particular form

$$F(R_{ij}) = 205 \left(-\frac{1}{R_{ij}^3} + \frac{1}{R_{ij}^5} \right).$$

For economy only, we assume next that two particles interact only within the local interaction distance $D = 1.3 \, \text{cm}$, which is the solution of the equation $\frac{dF}{dR_{ij}} = 0$. We use this criterion because σ is a molecular parameter, not a particle parameter, so that 2.5σ is not defined.

The dynamical equation of motion for each particle P_i of the system is then given by

$$\frac{d^2 \vec{R}_i}{dt^2} = -980\vec{\delta} + \frac{\alpha}{m} \sum_{\substack{j \\ j \neq i}} (205) \left[\frac{1}{R_{ij}^5} - \frac{1}{R_{ij}^3} \right] \frac{\vec{R}_{ji}}{R_{ij}}; \quad R_{ij} < D, \tag{6.10}$$

in which $\vec{\delta} = (0,1)$, α is a parameter, and $i = 1, 8479$. The reason for the introduction of the parameter α is that particle interaction should be local relative to gravity, that is, gravity must dominate for R_{ij} less than, but close to, unity, which we assume to mean for $R_{ij} \geq 0.7$. (Other possibilities are still under consideration.) Indeed, this is the case if we choose $\alpha = m$, since for $R_{ij} = 0.9, 0.8, 0.7$, and 0.6, F takes the values 66, 225, 622, and 1687, respectively, the first three of which are less than 980. Thus, the dynamical equations for the particles are

$$\frac{d^2 \vec{R}_i}{dt^2} = -980\, \vec{\delta} + \sum_{\substack{j \\ j \neq i}} (205) \left[\frac{1}{R_{ij}^5} - \frac{1}{R_{ij}^3} \right] \frac{\vec{R}_{ji}}{R_{ij}}; \quad R_{ij} < D. \tag{6.11}$$

6.3. Particle Equilibrium

We now allow the 8479 particles in Fig. 6.1 to find their own equilibrium when interacting in accordance with (6.11). We choose $\Delta t = 0.00005$ and use the usual reflection protocol, that is, if a particle crosses a wall, it is reflected back symmetrically, its velocity component tangent to the wall is set to zero, and its velocity component perpendicular to the wall is multiplied by -1. The initial motion of the system is almost one of free fall. So for the first 40000 time steps, each velocity is damped by the factor 0.2 every 20000 time steps. For the next 40000 time steps, each velocity is damped by the factor 0.4 every 20000 time steps. For the third 40000 time steps, each velocity is damped by the factor 0.7 every 20000 steps. For the final 40000 steps the damping is removed. In this fashion, the particles are eased down into the configuration shown in Fig. 6.2. Finally, to obtain a square set of particles, all particles with $y_i > 30$ in Fig. 6.2 are removed to yield the 7549 particle set shown in Fig. 6.3. The positions and velocities of these particles are used as the initial data for the cavity flow examples to be described next.

6.4. Examples

In each of the following examples $\Delta t = 0.00001$ and $J = 20000$. If a particle has crossed one of the lower three walls, its position is reflected back symmetrically, its velocity component parallel to the wall is set to zero, and its velocity component parallel to the wall is multiplied by -1. If the particle has crossed the moving wall, its position is reflected back symmetrically, its Y component of velocity is multiplied by -1, and its X component of velocity is increased by the wallspeed V.

Fig. 6.2.   8479 settled particles.

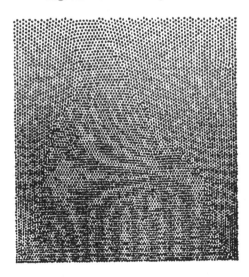

Fig. 6.3.   7459 particles in a square cavity.

Example 1. Let $V = -40$. Then Figs. 6.4–6.7 show the development of a primary vortex at the respective times $t = 0.2$, 0.6, 1.0, 1.4. Entirely similar results followed for $J = 14000$, 11000, 9000. Figure 6.7 compares most favorably with that in Fig. 2.6.

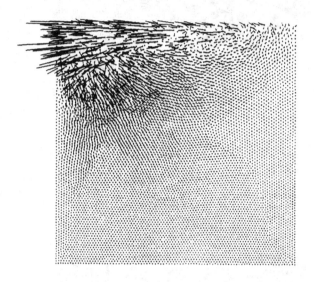

Fig. 6.4. $V = -40$, $t = 0.2$.

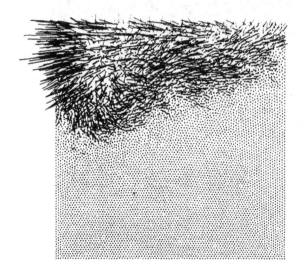

Fig. 6.5. $V = -40$, $t = 0.6$.

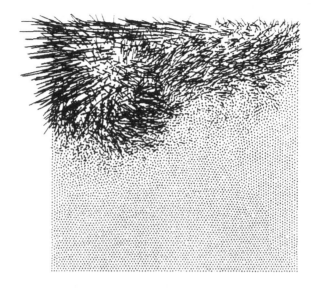

Fig. 6.6. $V = -40$, $t = 1.0$.

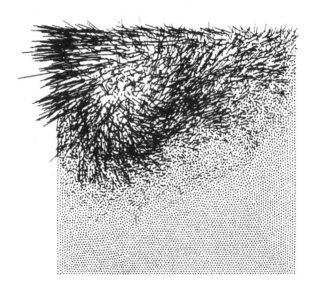

Fig. 6.7. $V = -40$, $t = 1.4$.

Example 2. Let $V = -100$. Figures 6.8–6.10 show the rapid development of a primary vortex, larger than that in Example 1, at the respective times $t = 0.2, 0.6, 1.0$. In this example, the velocity vectors had to be reduced by the factor $vm = 0.02$ for graphical clarity. Entirely similar results followed for $J = 14000, 11000, 9000$.

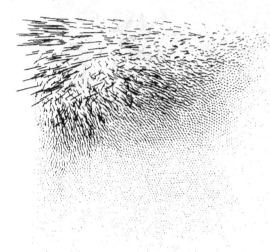

Fig. 6.8. $V = -100$, $t = 0.2$.

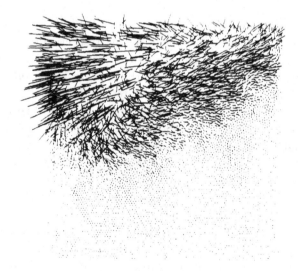

Fig. 6.9. $V = -100$, $t = 0.6$.

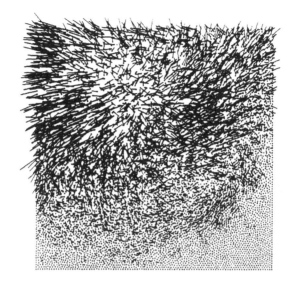

Fig. 6.10.　$V = -100$, $t = 1.0$.

6.5.　Turbulence

To simulate turbulence, set $V = -3000, \Delta t = 1.0(10)^{-6}, J = 40000$,
Figs. 6.11–6.13 show the flow development at the respective times 0.04,
0.08, 0.12. At $t = 0.12$, the time step had to be decreased to $5.0(10)^{-7}$
to avoid instability. Figure 6.14 shows the turbulent flow at $t = 0.15$ with
$J = 60000$, to make up for the decrease in time step. Entirely similar results
for Figs. 6.11–6.13 followed with $J = 30000, 20000, 10000$, while results
entirely similar to that in Fig. 6.14 followed for $J = 80000, 70000$, and
50000. In all the figures, the velocity vectors were decreased by $vm = 0.003$
for graphical clarity. Figure 6.12 compares most favorably with Fig. 2.14
while Fig. 6.14 compares most favorably with Fig. 2.16.

　　In searching for small vortices, attention was confined to within a circle
of radius 1 cm around each particle. There resulted at $t = 0.12$ 355 small
vortices which are shown in Fig. 6.15. Moreover, Fig. 6.16 shows the distribution of 349 small vortices at the time $t = 0.15$, so that, after only
0.03 sec, this figure shows the rapid change in much of the cavity of the
vortex distribution shown in Fig. 6.15.

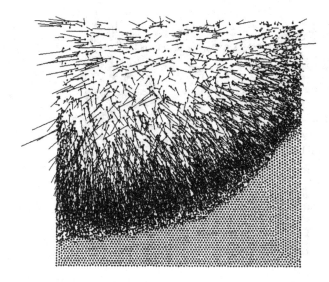

Fig. 6.11. $V = -3000$, $J = 40000$, $t = 0.04$.

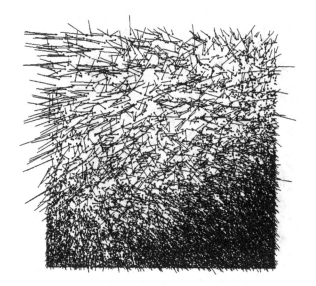

Fig. 6.12. $V = -3000$, $J = 40000$, $t = 0.08$.

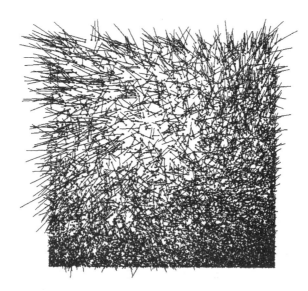

Fig. 6.13. $V = -3000$, $J = 40000$, $t = 0.12$.

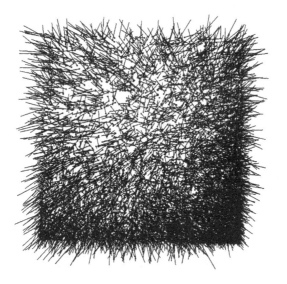

Fig. 6.14. $V = -3000$, $J = 60000$, $t = 0.15$.

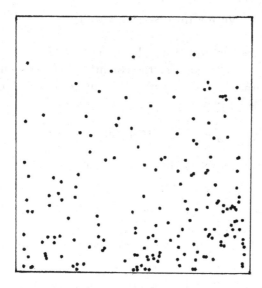

Fig. 6.15.   Small vortices (355) in Fig. 6.13.

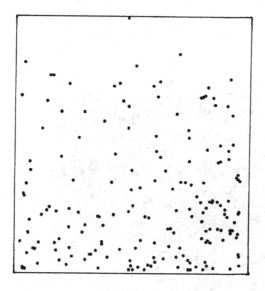

Fig. 6.16.   Small vortices (349) at $t = 0.15$ with $J = 60000$.

6.6. Heating Water Vapor in a Square Cavity

Since heating is of great interest in the large, let us begin now with the 7549 particle set in Fig. 6.3, but vary the conditions on the boundary in order to simulate heating on the bottom wall. A linear heating formula on the bottom wall is taken to be

$$v_y = vel(x + 15.0)/30,$$

in which *vel* is a parameter. Note that $v_y = 0$ at $x = -15.0$ and $v_y = vel$ at $x = 15.0$.

Example 1.   Set $vel = 600$. For $\Delta t = (10)^{-5}, J = 20000$, Figs. 6.17–6.20 show the development of a primary vortex at the respective times 0.2, 0.4, 0.6, 0.8. For graphical clarity, the velocity vectors have been reduced.

Example 2.   Set $vel = 3000$. The resulting flow is shown at 0.04, 0.08, 0.12 in Figs. 6.21–6.23. Figure 6.23 reveals the strong crosscurrent in the lower central area which is associated with the onset of turbulence. Figure 6.24 reveals that Fig. 6.23 has 609 small vortices and further calculations show that these appear and disappear quickly.

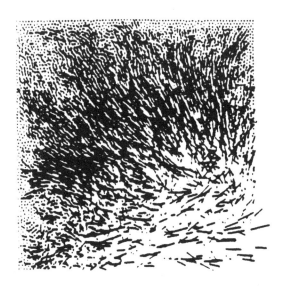

Fig. 6.17. $T = 2.0$.

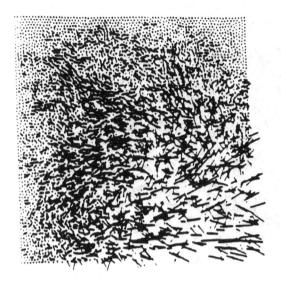

Fig. 6.18. $T = 4.0$.

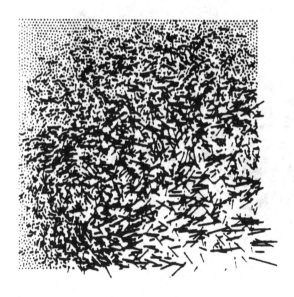

Fig. 6.19. $T = 8.0$.

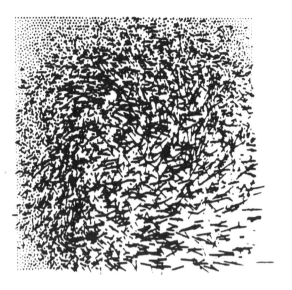

Fig. 6.20. $T = 14.0$.

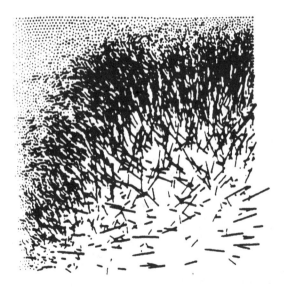

Fig. 6.21. $T = 0.3$.

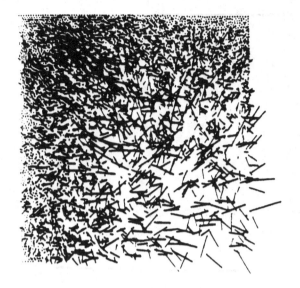

Fig. 6.22. $T = 0.9$.

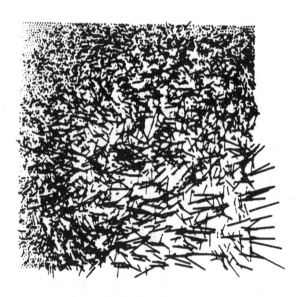

Fig. 6.23. $T = 1.5$.

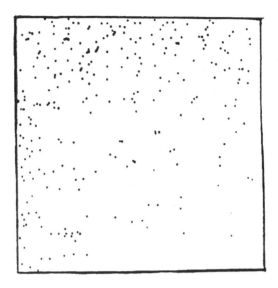

Fig. 6.24.   274 small vortices in Fig. 6.23.

6.7.   A Speculative Study of Liquid Water

In Sections 6.1–6.5, a particle mechanics approach was applied to study turbulence for water *vapor* in the context of the classical cavity problem. The primary aim was to develop a mechanism for turbulence. The primary aim now is to develop a turbulence mechanism in the large for *liquid* water, with the knowledge that no accurate Lennard–Jones potential is known for liquid water.

Let us choose first an approximate Lennard–Jones potential for water molecules. We choose the very same formula as for water vapor, that is,

$$\phi(r_{ij}) = (1.9646)10^{-13} \left[\frac{2.725^{12}}{r_{ij}^{12}} - \frac{2.725^{6}}{r_{ij}^{6}} \right] \text{erg} \ \left(\frac{\text{grcm}^2}{\text{sec}^2} \right) \qquad (6.12)$$

in which r_{ij} is measured in angstroms (Å). This choice is, of course, incorrect. We will however explore what results can still be obtained from use of this formula which neglects the hydrogen bonding of close water molecules. It should be observed also that there is no clear delination between dense water vapor and liquid water.

Note that our choice seems to be neither more nor less correct than potentials assumed by others. Rapaport's (1995) selection is the TIP4P potential, which has the form

$$\phi = \frac{q_\mu q_\nu e^2}{r_{i\mu,j\nu}} + \frac{A_{\mu\nu}}{r_{i\mu,j\nu}^{12}} - \frac{C_{\mu\nu}}{r_{i\mu,j\nu}^6}.$$

In this potential, *no* constants are specified. Indeed, as stated by Rapaport: "The charges associated with the sites, while maintaining some resemblance to the actual molecule, are generally regarded as parameters that can be adjusted to fit known molecular properties ..." In the Rahman and Stillinger references (1971, 1974), pointed to by Rapaport, the following unusual way is made to determine the Lennard–Jones portion of their potential. In their words (1971): "Since the water molecule and the neon atom are isoelectronic closed shell systems, the parameters ϵ and σ ... were chosen ... to be the accepted neon values. ..". These observations are made not to excuse our choice, but to indicate how poor all related studies are. We continue then only to see what tenuous results can be obtained if one neglects hydrogen bonding.

When one wishes to accumulate molecules into particles, problems arise if one wishes to consider *liquid* motion. The reason is that it is customary to consider liquids to be *incompressible*. In the case of water, we know that water is compressible because it conducts sound. Nevertheless, the compression waves which result can be of the order of 75 Å, so that an assumption of incompressibility to simulate liquid motions in the large becomes reasonable. We therefore aim at incompressibility as we continue.

The force $\vec{F}_{ij}$ exerted on P_i by P_j is then

$$\vec{F}_{ij} = (1.9646)10^{-5}\left[\frac{12(2.725^{12})}{r_{ij}^{13}} - \frac{6(2.725^6)}{r_{ij}^7}\right]\frac{\vec{r}_{ji}}{r_{ij}} \text{ dynes } \left(\frac{\text{grcm}}{\text{sec}^2}\right).$$

(6.13)

Recall that $F_{ij} = \|\vec{F}_{ij}\| = 0$ implies that $r_{ij} = 3.06$ Å, which is the equilibrium distance, and, in a regular triangle with edge $r_{ij} = 3.06$ Å, the altitude is 2.65 Å.

Our previous attempts to accumulate molecules into particles and then simulate liquid flow have centered on particle potentials of the form

$$\phi = -\frac{A}{R_{ij}^q} + \frac{B}{R_{ij}^p}, \quad A > 0, \ B > 0, \ p > q$$

with R_{ij} in cm and with the choices $(p, q) = (5, 3), (4, 2), (3, 1), (2, 1)$. In *every* case the resulting cavity flow exhibited large, and hence undesirable, compression waves. In this section we will apply a potential for which no observable compression waves will be apparent for the classical cavity problem.

For $vel = 0.0001$, with position given in centimeters and speed in centimeters per second, let

$$x(1) = -0.2, \quad y(1) = 0.0, \quad vx(1) = -vel, \quad vy(1) = vel$$
$$x(i + 1) = x(i) + 0.05, \quad y(i + 1) = 0.0, \quad vx(i + 1) = -vel,$$
$$vy(i + 1) = vel, \qquad i = 1, 8$$
$$x(10) = -0.175, \quad y(10) = 0.0433, \quad vx(10) = -vel, \quad vy(10) = -vel$$
$$x(i + 1) = x(i) + 0.05, \quad y(i + 1) = 0.0433, \quad vx(i + 1) = vel,$$
$$vy(i + 1) = -vel, \qquad i = 10, 16$$
$$x(i + 1) = x(i - 16), \quad y(i + 1) = y(i - 16) + 0.0866,$$
$$vx(i + 1) = -vx(i - 16), vy(i + 1) = -vy(i - 16), i = 17, 3500.$$

The triangulated particle arrangement lies in a rectangle which is 0.4 cm wide and approximately 17.8 cm high.

6.8.   Particle Equations of Motion

We assume for two particles P_i, P_j the potential

$$\phi(R_{ij}) = -\frac{A}{R_{ij}} + \frac{B}{(1.01)R_{ij}^{1.01}} \text{ erg}, \tag{6.14}$$

in which R_{ij} is the distance in centimeters between the particles. Thus, the force $\vec{F}_{ij}$ which particle P_j exerts on particle P_i is

$$\vec{F}_{ij} = \left(-\frac{A}{R_{ij}^2} + \frac{B}{R_{ij}^{2.01}} \right) \frac{\vec{R}_{ji}}{R_{ij}} \text{ dynes}. \tag{6.15}$$

Our first problem is to determine A and B. Let $F_{ij} = \|\vec{F}_{ij}\|$. Then in consistency with the way the particle positions were generated, let $F(0.05) = 0$, so that

$$-\frac{A}{(0.05)^2} + \frac{B}{(0.05)^{2.01}} = 0. \tag{6.16}$$

In order to determine a second equation, some relevant observations must be made first.

Note that the number N of *molecules* which can be arranged in the 0.4 cm by 17.6 cm rectangle using a regular triangular grid is

$$N = \frac{(0.4)(17.6)}{(3.06)(2.65)} 10^{16} = (0.86817)10^{16}. \tag{6.17}$$

Also, note that since the mass M of a water molecule is $(30.103)10^{-24}$ gr, the total mass of the water molecules inside the rectangle is

$$M = (26.13456)10^{-8} \, \text{gr}. \tag{6.18}$$

Distributing this mass over the 3500 particles for conservation of total mass yields an individual particle mass m of

$$m = (7.467)10^{-11}. \tag{6.19}$$

The total potential energy P_M of the molecular configuration is, approximately,

$$P_M = 3 \sum^{(0.86817)10^{16}} \left\{ (1.9646)10^{-13} \left[\left(\frac{2.725}{3.06}\right)^{12} - \left(\frac{2.725}{3.06}\right)^6 \right] \right.$$
$$= -(1.279)10^3. \tag{6.20}$$

In order to determine the kinetic energy K_M of the molecular system, we now fix the temperature at $15°C$, at which temperature the modal speed of a water molecule is $3.65 \, \text{Å/ps}$, or, equivalently, $3.65(10)^4 \, \text{cm/sec}$. Hence, the kinetic energy of the molecular system is

$$K_M = \frac{1}{2}(30.103)10^{-24}((3.65(10)^4))^2(0.86817)10^{16} = 174 \, \text{erg}.$$

Thus, from (6.20), the total energy E_M of the molecular system is

$$E_M = -1279 + 174 = -1105 \, \text{erg}. \tag{6.21}$$

On the other hand, the total potential energy E_m of the particle configuration is, approximately,

$$E_m = 3(3500) \left[-\frac{A}{0.05} + \frac{B}{(1.01)(0.05)^{1.01}} \right],$$

and by our choice of $vel = 0.0001$, the kinetic energy of the particle system is negligible. Thus, we set E_M equal to E_m and find that the second equation for A and B is

$$-20A + 20.40417B = -0.1052. \tag{6.22}$$

The solution of (6.16) and (6.22) is $A = 0.5318, B = 0.5161$. Thus,

$$\vec{F}_{ij} = \left(-\frac{0.5318}{R_{ij}^2} + \frac{0.5161}{R_{ij}^{2.01}} \right) \frac{\vec{R}_{ji}}{R_{ij}}. \tag{6.23}$$

The dynamical equation of motion for each particle P_i of the system is then given by

$$\frac{d^2 \vec{R}_i}{dt^2} = -980\vec{\delta} + \frac{\alpha}{m} \sum_j \left[-\frac{0.5318}{R_{ij}^2} + \frac{0.5161}{R_{ij}^{2.01}} \right] \frac{\vec{R}_{ji}}{R_{ij}}, \tag{6.24}$$

in which α is a parameter and $\vec{\delta} = (0, 1)$. We choose $\alpha = 100\,m/5$, the convenience of which will be described shortly. Hence, (6.24) reduces to

$$\frac{d^2 \vec{R}_i}{dt^2} = -980\vec{\delta} + \sum_j \left[-\frac{10.636}{R_{ij}^2} + \frac{10.322}{R_{ij}^{2.01}} \right] \frac{\vec{R}_{ji}}{R_{ij}}. \tag{6.25}$$

Note now that

$$-\frac{10.636}{(0.01235)^2} + \frac{10.322}{(0.01235)^{2.01}} = 981,$$

so that for $R_{ij}^* = 0.01235$ the local force balances the force of gravity. As is common in classical molecular mechanics, we now choose a local distance of interaction D to be $4R_{ij}^* = 0.0494$, so that (6.25) is now modified to

$$\frac{d^2 \vec{R}_i}{dt^2} = -980\vec{\delta} + \sum_j \left[-\frac{10.636}{R_{ij}^2} + \frac{10.322}{R_{ij}^{2.01}} \right] \frac{\vec{R}_{ji}}{R_{ij}}, \quad R_{ij} < D. \qquad (6.26)$$

Note also that

$$-\frac{10.636}{(0.0494)^2} + \frac{10.322}{(0.0494)^{2.01}} = -4358 + 4359 = 1,$$

so that if $R_{ij} = 0.0494$, gravity dominates and the local force is negligible. It is interesting to observe that 0.0494 is approximately 0.05, which was the grid size in the generation of the set of 3500 particles.

Finally, for computational convenience, we make the change of variables $T = (10)^{\frac{1}{2}} t$, so that (6.26) reduces finally to

$$\frac{d^2 \vec{R}_i}{dT^2} = -98\vec{\delta} + \sum_j \left[-\frac{1.0636}{R_{ij}^2} + \frac{1.0322}{R_{ij}^{2.01}} \right] \frac{\vec{R}_{ji}}{R_{ij}}, \quad R_{ij} < D. \qquad (6.27)$$

6.9. Particle Equilibrium

We now allow the 3500 particles to find their own equilibrium when interacting in accordance with (6.27). Initially, we choose $\Delta T = 0.0001$ and use the usual reflection protocol.

The initial motion of the system is almost one of free fall. So, for the first 7700000 steps, the velocity of each particle was damped at each time step by the factor 0.6. For the next five sets of 350000 time steps, the damping factor was changed, respectively, to 0.9, 0.95, 0.98, 0.995, and 0.999. For a final 350000 time steps, the damping was removed. The resulting stable configuration is shown in Fig. 6.25. From Fig. 6.25 we now extract the relatively dense set of 1718 particles in range $-0.125 \le x \le 0.125, 0 \le y \le 0.25$, shown in Fig. 6.26. This eliminates the surface tension particles at the top of Fig. 6.25, which are of no consequence, since these particles play no role in the cavity problem.

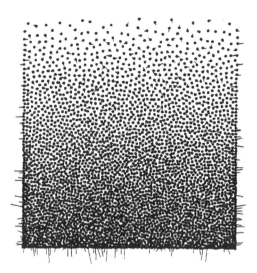

Fig. 6.25.   Stable water configuration.

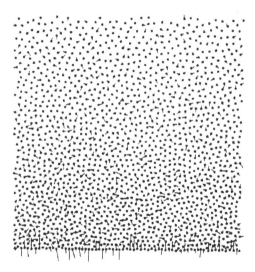

Fig. 6.26.   A dense subset of Fig. 6.25.

6.10.   Primary Vortex Generation

In this section we consider the generation of primary vortices for the cavity problem. In each example, it is to be observed that the fluid motion is counterclockwise and the results are in agreement with both experiment and numerical calculation with the Navier–Stokes equations (Schlichting (1960), Roache (1972)).

The computations proceed under the same considerations as those discussed in Section 2.4.

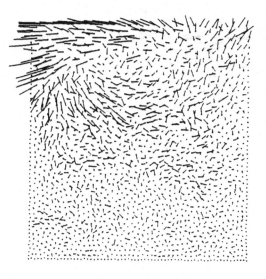

Fig. 6.27. $V = -1$, $J = 120000$, $T = 0.32$.

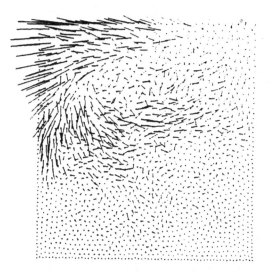

Fig. 6.28. $V = -1$, $J = 120000$, $T = 0.64$.

Example 1.   Consider now the cavity problem for the 1718 particles in
Fig. 6.26. Let $V = -1.0, \Delta T = 0.000002, J = 120000$. Figures 6.27–6.30
show the resulting motion at the times $T = 0.32$, 0.64, 0.96, 1.28. Similar
results were found for $J = 160000$, 80000, 40000. For clarity in the figures,
the velocities were decreased by the factor $mv = 0.021$.

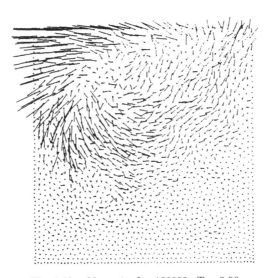

Fig. 6.29. $V = -1$, $J = 120000$, $T = 0.96$.

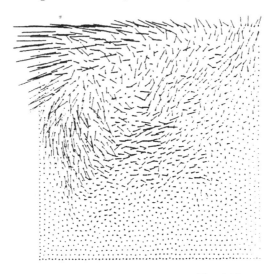

Fig. 6.30. $V = -1$, $J = 120000$, $T = 1.28$.

Example 2. Example 1 is repeated with the change $V = -2.0$. Figures 6.31–6.34, at the respective times $T = 0.32, 0.64, 0.96, 1.28$, show the development of a primary vortex which is larger than the one developed in Example 1. Similar results were found for $J = 160000, 80000, 40000$.

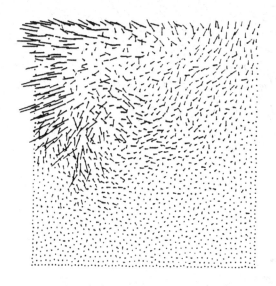

Fig. 6.31. $V = -2$, $J = 120000$, $T = 0.32$.

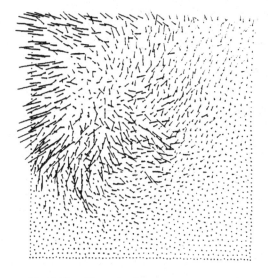

Fig. 6.32. $V = -2$, $J = 120000$, $T = 0.64$.

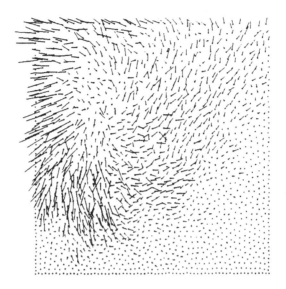

Fig. 6.33. $V = -2$, $J = 120000$, $T = 0.96$.

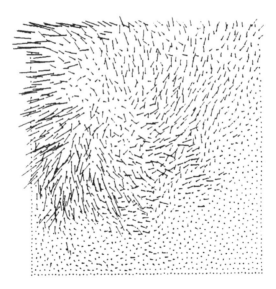

Fig. 6.34. $V = -2$, $J = 120000$, $T = 1.28$.

Example 3. Example 1 is repeated with the changes $V = -4.0$, $J = 80000$. Figures 6.35–6.38, at the respective times $T = 0.32$, 0.64, 0.96, 1.28, show the development of a primary vortex which is larger than that in Example 2. Similar results followed for $J = 160000$, 120000, 40000. In

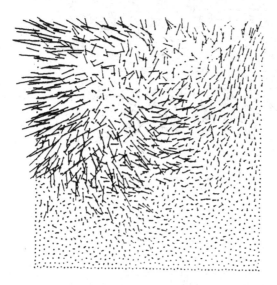

Fig. 6.35. $V = -4$, $J = 80000$, $T = 0.32$.

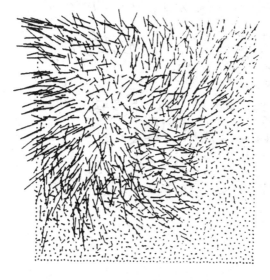

Fig. 6.36. $V = -4$, $J = 80000$, $T = 0.64$.

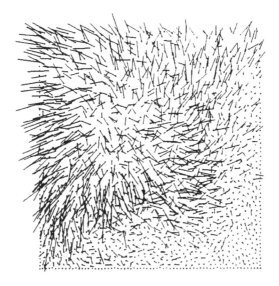

Fig. 6.37. $V = -4$, $J = 80000$, $T = 0.96$.

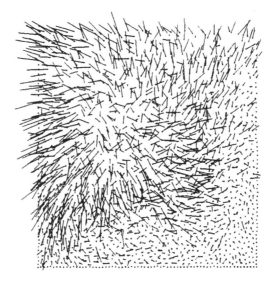

Fig. 6.38. $V = -4$, $J = 80000$, $T = 1.28$.

Figs. 6.35–6.38, the velocity vectors had to be decreased by the factor $mv = 0.0095$ for graphical clarity. Similar results were found for $J = 160000$, 120000, 40000.

Example 4. Example 1 was repeated with the changes $V = -20.0, \Delta T = 0.000001$. No primary vortex resulted. We then turn to this particular case and study it in detail next.

6.11. Turbulence

To simulate turbulence, set $V = -20, \Delta T = 0.000001$. Figures 6.39–6.42 show the flow development at $T = 0.16, 0.32, 0.48, 0.64$, for $J = 40000$. Each figure reveals the development of a strong current across the usual counterclockwise motion of primary vortices. For graphical clarity velocity vectors had to be decreased by the factor $mv = 0.0055$. Similar results followed for $J = 160000, 120000, 80000$.

Figure 6.43 shows an enlargement of the right portion of the flow in Fig. 6.42 in the region $0.11 < x < 0.125$, and reveals the crosscurrent readily.

Again, to support the contention that Figs. 6.39–6.42 represent fully turbulent flow, we now recall the definition of a small vortex. For $3 \leq M \leq 6$,

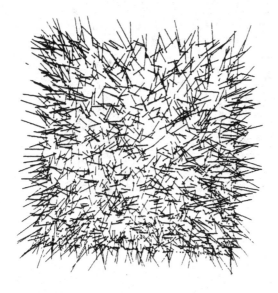

Fig. 6.39. $V = -20$, $J = 40000$, $T = 0.16$.

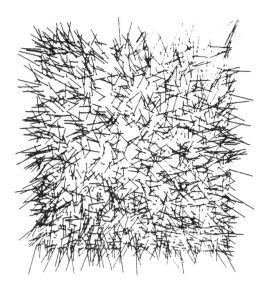

Fig. 6.40. $V = -20$, $J = 40000$, $T = 0.32$.

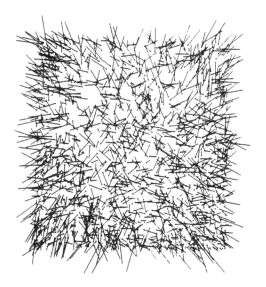

Fig. 6.41. $V = -20$, $J = 40000$, $T = 0.48$.

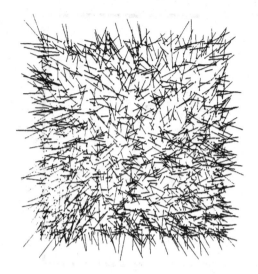

Fig. 6.42. $V = -20$, $J = 40000$, $T = 0.64$.

Fig. 6.43.   The section of Fig. 6.42 in the range $0.11 \leq x \leq 0.125$.

we define a *small vortex* as a flow in which M molecules nearest to an $(M + 1)$st molecule rotate either clockwise or counterclockwise about the $(M + 1)$st molecule and, in addition, the $(M + 1)$st molecule lies *interior* to a simple polygon determined by the given M molecules. In searching for

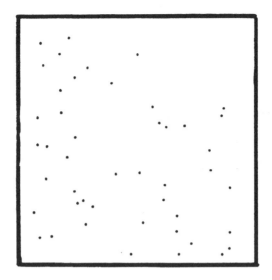

Fig. 6.44.   87 small vortices in Fig. 6.41.

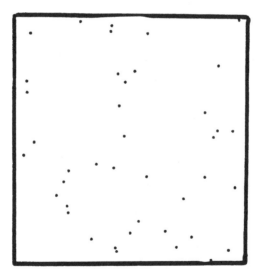

Fig. 6.45.   80 small vortices in Fig. 6.42.

small vortices, attention was confined to within a circle of radius 0.02 cm around each particle. Figure 6.44 shows 87 small vortices for Fig. 6.41, while Fig. 6.45 shows 80 small vortices for Fig. 6.42. Figures 6.44 and 6.45 are entirely different. Thus, in a change of only $\Delta T = 0.16$, the number and the distribution of the small vortices has changed completely, as is characteristic of turbulent flow.

6.12. The Fortran Program PARTICLE.FOR

In this section we give a typical program used in this chapter.

FORTRAN PROGRAM PARTICLE.FOR

```
C  Total number of particles is 7549.
C  Exponent parameters are 3 and 5.
        DOUBLE PRECISION XO(7549),YO(7549),VXO(7549),VYO(7549),
       1X(7549,2),Y(7549,2),VX(7549,2),VY(7549,2),
       1FX(7549),FY(7549),ACX(7549),ACY(7549),
       1F(7549),R2,R4
        OPEN (UNIT=21,FILE='PARTICLE.DAT',STATUS='OLD')
        OPEN (UNIT=31,FILE='PARTICLE.OUT',STATUS='NEW')
        OPEN (UNIT=71,FILE='PARTICLE.OUT1',STATUS='NEW')
        OPEN (UNIT=72,FILE='PARTICLE.OUT2',STATUS='NEW')
        OPEN (UNIT=73,FILE='PARTICLE.OUT3',STATUS='NEW')
        OPEN (UNIT=74,FILE='PARTICLE.OUT4',STATUS='NEW')
        K=1
        WALLSPEED= - 40.0
        READ (21,10)  (XO(I),YO(I),VXO(I),VYO(I),I=1,7549)
10      FORMAT (4F12.6)
        DO 30 I=1,7549
        X(I,1)=XO(I)
        Y(I,1)=YO(I)
        VX(I,1)=VXO(I)
        VY(I,1)=VYO(I)
30      CONTINUE
        GO TO 3456
C Cycle.
65      DO 70 I=1,7549
        X(I,1)=X(I,2)
        Y(I,1)=Y(I,2)
        VX(I,1)=VX(I,2)
        VY(I,1)=VY(I,2)
```

```
70    CONTINUE
C     Calculate forces and accelerations using an accumulator.
      DO 701 I=1,7549
      ACX(I)=0.0
      ACY(I)=0.0
      F(I)=0.0
701   CONTINUE
3456  DO 78 I=1,7548
      ACXI=ACX(I)
      ACYI=ACY(I)
      XI=X(I,1)
      YI=Y(I,1)
      IP1=I+1
      DO 77 J=IP1,7549
      R2=(XI-X(J,1))**2+(YI-Y(J,1)**2
C The local interaction distance D is 1.3.
      IF (R2.GE.1.69) GO TO 9000
      R4=R2*R2
      F(J)=(-207.+207./(R2))/(R4)
      GO TO 9001
9000  F(J)=0.0
9001  ACX(J)=ACX(J)-F(J)*(XI-X(J,1))
      ACY(J)=ACY(J)-F(J)*(YI-Y(J,1))
      ACXI=ACXI+F(J)*(XI-X(J,1))
      ACYI=ACYI+F(J)*(YI-Y(J,1))
77    CONTINUE
      ACX(I)=ACXI
      ACY(I)=ACYI
78    CONTINUE
C  Insert gravity.
      DO 5634 I=1,7549
      ACY(I)=ACY(I)-980.
5634    CONTINUE
C Calculate new positions and velocities for wallspeed −40.
      DO 7123 I=1,7549
      VX(I,2)=VX(I,1)+0.00001*ACX(I)
      VY(I,2)=VY(I,1)+0.00001*ACY(I)
```

```
      X(I,2)=X(I,1)+0.00001*VX(I,2)
      Y(I,2)=Y(I,1)+0.00001*VY(I,2)
7123  CONTINUE
C At this point  insert wall reflection.
      DO 995 I=1,7549
      IF (X(I,2).LT.15.0) GO TO 992
      X(I,2)=30.-X(I,2)
      VX(I,2)=-1.0*VX(I,2)
      VY(I,2)=0.0
992   IF (X(I,2).GT.-15.0) GO TO 993
      X(I,2)=-30.-X(I,2)
      VX(I,2)=-1.0*VX(I,2)
      VY(I,2)=0.0
993   IF (Y(I,2).GT.0.0) GO TO 994
      Y(I,2)=-Y(I,2)
      VX(I,2)=0.0
      VY(I,2)=-1.0*VY(I,2)
994   IF (Y(I,2).LT.30.0) GO TO 995
      Y(I,2)=60.-Y(I,2)
      VX(I,2)=VX(I,2)+WALLSPEED
      VY(I,2)=-1.0*VY(I,2)
995   CONTINUE
      K=K+1
82    IF (K.LT.10000) GO TO 65
      WRITE (31,10) (X(I,2),Y(I,2),VX(I,2),VY(I,2),I=1,7549)
      STOP
      END
```

Chapter 7

The Flow of Water Vapor Around
a Flat Plate

7.1. Introduction

In this chapter we will study the planar flow of water vapor molecules around a flat plate. Flows around plates in the large have been studied extensively by fluid dynamicists (see, e.g. M. Van Dyke, (1982)). The flow which we will explore is on the nano level.

7.2. Mathematical and Physical Preliminaries

Let us recall first an approximate Lennard–Jones potential for water vapor molecules, that is,

$$\phi(r_{ij}) = (1.9646)10^{-13} \left[\frac{2.725^{12}}{r_{ij}^{12}} - \frac{2.725^6}{r_{ij}^6} \right] \text{erg} \quad \left(\frac{\text{grcm}^2}{\text{sec}^2} \right) \qquad (7.1)$$

in which r_{ij} is measured in angstroms (Å). The force $\vec{F}_{ij}$ exerted on P_i by P_j is then

$$\vec{F}_{ij} = (1.9646)10^{-5} \left[\frac{12(2.725^{12})}{r_{ij}^{13}} - \frac{6(2.725^6)}{r_{ij}^7} \right] \frac{\vec{r}_{ji}}{r_{ij}} \text{dynes} \quad \left(\frac{\text{grcm}}{\text{sec}^2} \right).$$

$$(7.2)$$

From (7.2), it follows that the equilibrium distance between two molecules is 3.05 Å.

7.3.   Approximate Equations

From the discussion in Chapter 4, it follows that the equation of motion for a single water vapor molecule P_i acted on by a single water vapor molecule $P_j, i \neq j$, is

$$m_i \vec{a}_i = (1.9646)10^{-5} \left[\frac{12(2.725^{12})}{r_{ij}^{13}} - \frac{6(2.725^6)}{r_{ij}^7} \right] \frac{\vec{r}_{ji}}{r_{ij}}. \qquad (7.1)$$

Since the mass of a water molecule is $(30.103)10^{-24}\,\text{gr}$, Eq. (7.1) is equivalent to

$$\vec{a}_i = (160.33)10^{19} \left[\frac{818.90}{r_{ij}^{13}} - \frac{1}{r_{ij}^7} \right] \frac{\vec{r}_{ji}}{r_{ij}} \quad \left(\frac{\text{cm}}{\text{sec}^2} \right). \qquad (7.2)$$

For computational convenience, we rewrite Eq. (7.2) in $\text{Å}/(\text{ps}^2)$, to yield

$$\vec{a}_i = (160330) \left[\frac{818.90}{r_{ij}^{13}} - \frac{1}{r_{ij}^7} \right] \frac{\vec{r}_{ji}}{r_{ij}} \quad \left(\frac{\text{Å}}{\text{ps}^2} \right). \qquad (7.3)$$

On the molecular level, however, the effective force on P_i is *local*, that is it is determined only by close molecules, by which we will mean only those molecules within the distance $D = 3.39\,\text{Å}$.

From (7.3), then, the dynamical equation for water molecule P_i will be

$$\frac{d^2 \vec{r}_i}{dt^2} = (160330) \sum_{\substack{j \\ j \neq i}} \left[\frac{818.90}{r_{ij}^{13}} - \frac{1}{r_{ij}^7} \right] \frac{\vec{r}_{ji}}{r_{ij}}; \quad r_{ij} < D. \qquad (7.4)$$

The equations of motion for a system of water vapor molecules are then

$$\frac{d^2 \vec{r}_i}{dt^2} = (160330) \sum_{\substack{j \\ j \neq i}} \left[\frac{818.90}{r_{ij}^{13}} - \frac{1}{r_{ij}^7} \right] \frac{\vec{r}_{ji}}{r_{ij}}; \quad i = 1, 2, 3, \dots, N; \ r_{ij} < D.$$

$$(7.5)$$

Again, gravity is neglected.

7.4. Problem Formulation

Unless otherwise specified, the following general problem will be studied.

We first generate 49756 water molecules using the following algorithm:

$$x(1) = y(1) = 0.0$$
$$x(i+1) = x(i) + 3.05, \quad y(i+1) = 0.0, \qquad i = 1,260$$
$$x(262) = 1.525, \quad y(262) = 2.64$$
$$x(i+1) = x(i) + 3.05, \quad y(i+1) = 2.64, \qquad i = 262,520$$
$$x(i) = x(i-521), \quad y(i) = y(i-521) + 5.28, \qquad i = 522,49756.$$

We would like to replace various molecules by a rigid plate which is at a 45° angle in the basin. To do this conveniently we first rotate the particles generated above 45° by the following algorithm:

$$x^*(i) = 0.965925826289068x(i) + 0.258819045102521y(i), \quad i = 1,49756$$
$$y^*(i) = 0.965925826289068y(i) - 0.258819045102521x(i), \quad i = 1,49756.$$

Next we decrease the number of molecules to 12420 by choosing from the above set only those which lie in the rectangle $129 \leq x \leq 529, 0 \leq y \leq 250$. Finally we translate all these by $x(i) = x^*(i) - 329$, so that the 12420 molecules lie in the rectangle $-200 \leq x \leq 200, 0 \leq y \leq 250$. This result is shown in Fig. 7.1.

Next, we insert a 278 particle plate with a 45° slope between the two points $(-13.8673, 109.6126)$ and $(13.8327, 137.3127)$ and delete all molecules which are immediately adjacent to the plate and above it. This is shown in Fig. 7.2, which has 12394 vapor molecules and 278 fixed plate particles.

We now have the basin with the desired plate and are ready to simulate flow around the plate. For this reason we now assign to each molecule to the left of the plate a speed V in the x direction. Thus we now have the configuration shown in Fig. 7.3. The top, that is $y = 250$, and bottom that is, $y = 0$, of this figure will be taken as fixed walls. The region to the right of the plate will be called the shadow region.

Our problem is to describe the resulting flow around the plate.

For time step Δt (ps), and $t_k = k\Delta t$, two problems must be considered relative to the computations. The first problem is to prescribe a protocol when, computationally, a molecule has crossed a fixed wall into the exterior of the cavity. For each of the upper and lower walls, we will proceed with

Fig. 7.1.   12420 points.

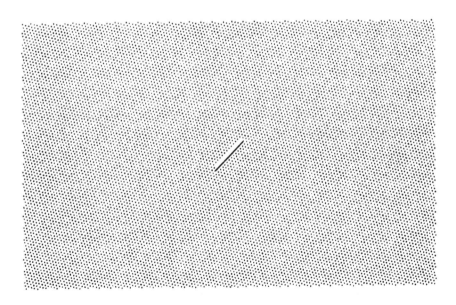

Fig. 7.2.   12672 points.

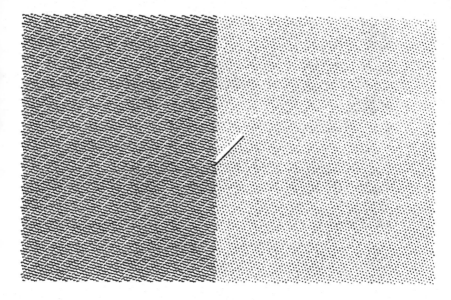

Fig. 7.3.   Initial velocities to left of the plate.

the no slip condition. If the molecule has crossed the plate, then it will be reflected back symmetrically and its x and y components of velocity will be replaced by its y and x components.

In order to better interpret gross fluid motion, we again introduce average velocities. For J a positive integer, let particle P_i be at $(x(i,k), y(i,k))$ at t_k and at $(x(i, k - J), y(i, k - J))$ at t_{k-J}. Then the average velocity $\vec{v}_{i,k,J}$ of P_i at t_k is defined by

$$\vec{v}_{i,k,J} = \left(\frac{x(i,k) - x(i, k - J)}{J\Delta t}, \frac{y(i,k) - y(i, k - J)}{J\Delta t} \right). \qquad (7.6)$$

In the examples to be described, we will discuss results for various values of J.

7.5.   Examples

Example 1.   For our first example let $V = 35, \Delta t = 4(10)^{-6}, J = 60000$. Then Figs. 7.4–7.12 show the resulting laminar flow at the respected times $T = 0.36, 0.60, 0.84, 1.08, 1.32, 1.56, 1.80, 2.04, 2.28$. The most noticeable aspect of the flow is its movement below the plate and then up into the shadow region. Figure 7.13 shows the instantaneous Brownian motion of

Fig. 7.4. $V = 35$, $T = 0.36$.

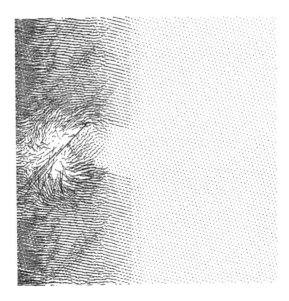

Fig. 7.5. $V = 35$, $T = 0.60$.

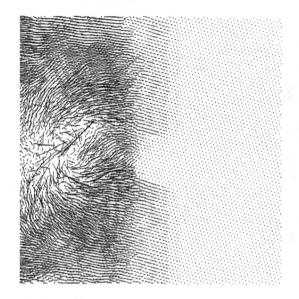

Fig. 7.6. $V = 35$, $T = 0.84$.

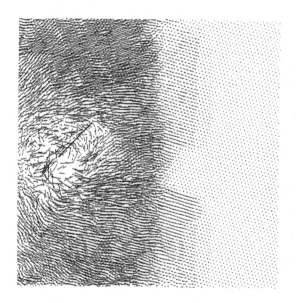

Fig. 7.7. $V = 35$, $T = 1.08$.

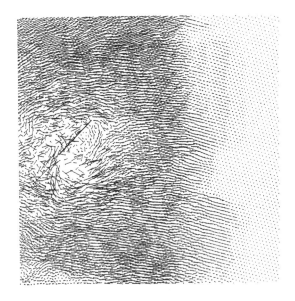

Fig. 7.8. $V = 35$, $T = 1.32$.

Fig. 7.9. $V = 35$, $T = 1.56$.

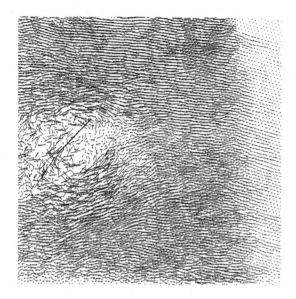

Fig. 7.10. $V = 35,\ T = 1.80.$

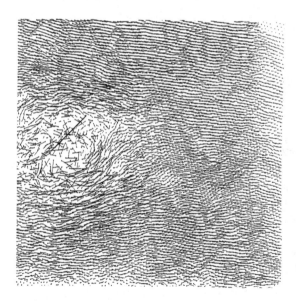

Fig. 7.11. $V = 35,\ T = 2.04.$

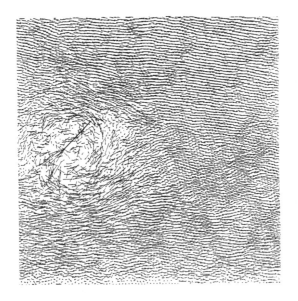

Fig. 7.12. $V = 35$, $T = 2.28$.

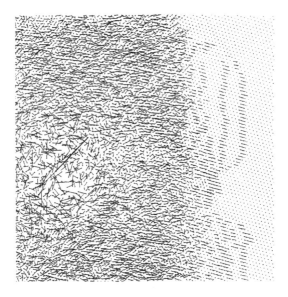

Fig. 7.13.   Brownian motion for $V = 35$, $T = 1.32$.

the flow at $T = 1.32$ and should be compared with Fig. 7.8. Similar results were obtained for $J = 30000, 90000$.

Example 2. Example 1 was modified by setting $V = 20, \Delta t = 8(10)^{-6}$. The flow was entirely similar to that of Example 1 but developed more slowly. Figure 7.14 shows the flow at $T = 2.16$ and should be compared to the flows shown in Figs. 7.7 and 7.8 at times $T = 1.08, 1.32$. The vectors in Fig. 7.14 should be smaller than those in Figs. 7.7 and 7.8, but have been rescaled for clarity.

Example 3. Example 1 was modified by setting $V = 55$. The flow was entirely similar to that of Example 1, but developed more rapidly. Figure 7.15 shows the flow at $T = 1.84$ and should be compared with Figs. 7.11 and 7.12.

Example 4. Example 1 was modified by setting $V = 500, \Delta t = 4(10)^{-7}, J = 120000$. Figure 7.16 shows the flow at the early time $T = 0.284$. The flow is entirely similar to that in Example 1 but has developed more rapidly. It should be compared with Fig. 7.10 at $T = 1.80$.

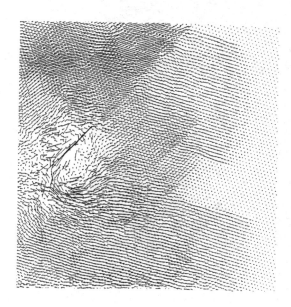

Fig. 7.14. $V = 20$, $T = 2.16$.

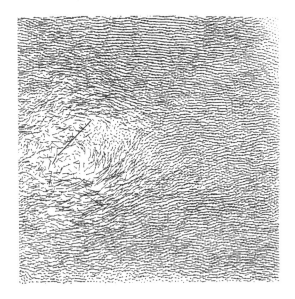

Fig. 7.15. $V = 55$, $T = 1.84$.

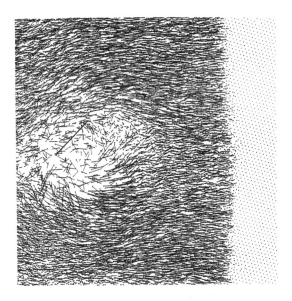

Fig. 7.16. $V = 500$, $T = 0.284$, $J = 120000$.

Example 5.   Example 1 was modified by setting $V = 6000, \Delta t = 4(10)^{-8}, J = 180000$. Figure 7.17 shows the flow at $4.8(10)^{-7}$. The velocity field has been rescaled extensively for graphical clarity. The flow now differs from all previous examples in that there is a strong current below the plate which crosses the usual laminar flow direction. Now, physically, a viewer does not see the velocity field in Fig. 7.17. The viewer sees in the area below the plate the rapid appearance and disappearance of small vortices. The 12 vortices in the range $-20 < x < 20, 40 < y < 80$ are

$$(-14.47, 44.25), (-15.53, 41.92), (-8.69, 49.93), (4.77, 46.52),$$
$$(14.31, 41.57), (-12.93, 57.97), (4.93, 54.35), (1.14, 58.19),$$
$$(5.89, 62.07), (-5.36, 76.55), (7.16, 67.10), (-6.13, 66.99)$$

at $4.8(10)^{-7}$. At $4.4(10)^{-7}$ there are 11 vortices, namely,

$$(-17.98, 41.21), (3.07, 43.28), (11.36, 41.98), (12.25, 50.78),$$
$$(1.76, 49.76), (9.06, 47.83), (-12.90, 58.84), (14.03, 56.63),$$
$$(13.03, 53.58), (-19.41, 70.77), (0.91, 46.48).$$

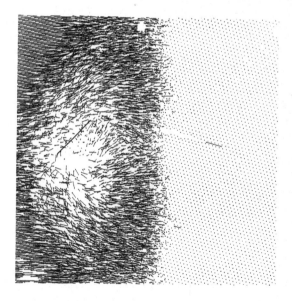

Fig. 7.17. $V = 6000$, $T = 4.8(10)^{-7}$, $J = 180000$.

Thus, in only $4(10)^{-8}$ sec, the vortices have changed in both number and in position. Note also that completely analogous results followed for $J = 240000$, $J = 210000$, $J = 150000$, $J = 120000$.

Example 6.   Example 5 was modified by setting $V = 3000$. The results were similar to those of Example 5 but were not as sharply defined.

7.6.   Remarks

Examples entirely analogous to those in Section 5 were repeated, but for air rather than for water vapor. The results were similar, but not identical. Figures 7.18–7.21 show the results for $V = 500$, $J = 120000$, at the respective times $T = 0.159, 0.223, 0.294, 0.366.$ Typically, the flow for air is more dilute immediately behind the plate than for water vapor. Figure 7.20 should be compared with Fig. 7.16.

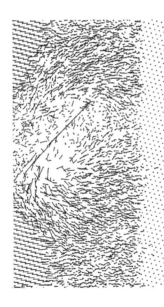

Fig. 7.18. $V = 500$, $T = 0.159$, $J = 120000$.

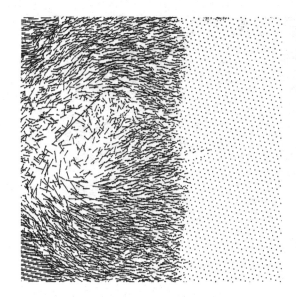

Fig. 7.19. $V = 500$, $T = 0.223$, $J = 120000$.

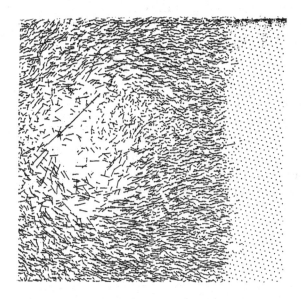

Fig. 7.20. $V = 500$, $T = 0.294$, $J = 120000$.

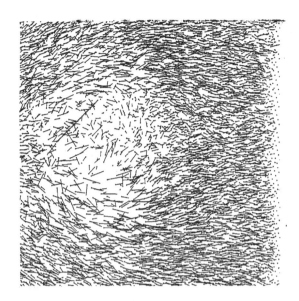

Fig. 7.21. $V = 500$, $T = 0.366$, $J = 120000$.

7.7.   The Fortran Program **PLATE.FOR**

We now provide a typical program used for the examples in Section 7.5.

FORTRAN PROGRAM PLATE.FOR

```
c This is for water vapor.
c TotaL number of particles is  12672.
c The moving particles are 1-12394, fixed are 12395-12672.
      double precision xo(12672),yo(12672),vxo(12672),vyo(12672),
     1x(12672,2),y(12672,2),vx(12672,2),vy(12672,2),
     1acx(12672),acy(12672)
      open (unit=21,file='adam7.dat',status='old')
      open (unit=31,file='adam7.out',status='new')
      k=1
      kk=1
      kprint=10000
      read (21,10) (xo(i),yo(i),vxo(i),vyo(i),i=1,12672)
10    format (4f12.4)
      do 30 i=1,12672
      x(i,1)=xo(i)
      y(i,1)=yo(i)
```

```
      vx(i,1)=vxo(i)
      vy(i,1)=vyo(i)
30    continue
      Go to 3456
65    do 70 i=1,12672
      x(i,1)=x(i,2)
      y(i,1)=y(i,2)
      vx(i,1)=vx(i,2)
      vy(i,1)=vy(i,2)
70    continue
      do 701 i=1,12394
      acx(i)=0.
      acy(i)=0.
      f=0.
701   continue
3456  do 78 i=1,12392
      acxi=acx(i)
      acyi=acy(i)
      xi=x(i,1)
      yi=y(i,1)
      ip1=i+1
      do 77 j=ip1,12394
      r2=(xi-x(j,1))**2+(yi-y(j,1))**2
c In this program the local interaction distance is 2.25sigma.
      if (r2.gt.46.) go to 9000
      r4=r2*r2
      r8=r4*r4
      f=160330.*((818.90/(r2*r4*r8))-1./(r8))
      go to 9001
9000  f=0.0
9001  acx(j)=acx(j)-f*(xi-x(j,1))
      acy(j)=acy(j)-f*(yi-y(j,1))
      acxi=acxi+f*(xi-x(j,1))
      acyi=acyi+f*(yi-y(j,1))
77    continue
      acx(i)=acxi
      acy(i)=acyi
78    continue
      do 7123 i=1,12394
```

```
        vx(i,2)=vx(i,1)+0.000004*acx(i)
        vy(i,2)=vy(i,1)+0.000004*acy(i)
        x(i,2)=x(i,1)+0.000004*vx(i,2)
        y(i,2)=y(i,1)+0.000004*vy(i,2)
7123    continue
c At this point we insert reflection.
        do 995 i=1,12394
        if (y(i,2).gt.250.) go to 900
        if (y(i,2).lt.0.0) go to 901
        if (y(i,1).gt.137.3125.and.y(i,2).gt.137.3125) go to 995
        if (y(i,1).lt.109.6126.and.y(i,2).lt.109.6126) go to 995
        if (x(i,1).lt.-13.8673.and.x(i,2).lt.-13.8673) go to 995
        if (x(i,1).gt.13.8327.and.x(i,2).gt.13.8327) go to 995
        if (y(i,1).ge.x(i,1)+123.4799.and.y(i,2).le.x(i,2)+123.4799) go to 902
        go to 995
900     y(i,2)=500.-y(i,2)
        vx(i,2)=0.0
        vy(i,2)=-vy(i,2)
        go to 995
901     y(i,2)=-y(i,2)
        vx(i,2)=0.0
        vy(i,2)=-vy(i,2)
        go to 995
902     x(i,2)=y(i,2)-123.4799
        y(i,2)=x(i,2)+123.4799
        vx(i,2)=vy(i,2)
        vy(i,2)=vx(i,2)
995     continue
        do 666 i=12395,12672
        x(i,2)=xo(i)
        y(i,2)=yo(i)
        vx(i,2)=0.0
        vy(i,2)=0.0
666     continue
        k=k+1
82      if (k.lt.30000) go to 65
        write (31,10) (x(i,2),y(i,2),vx(i,2),vy(i,2),i=1,12672)
        stop
        end
```

Chapter 8

Extant Problems with Continuum Models

8.1. Introduction

The last 350 years of continuum modelling has resulted in exceptional scientific understanding and achievement. Unfortunately, it has also lulled many of us into the belief that continuum models are the best models one can use and that these are immune from defects. Our limited world view too often has overlooked deficiencies of continuum models, a few of which will be discussed, for completeness, in this chapter.

8.2. Concepts of Infinity

The basis of all continuum modelling is the concept of *infinity*, to which we turn our attention first.

There are basically two concepts of infinity. The *first* is characterized by the symbol ∞ and denotes a *process*, that is, the process of growing beyond all bound. It is *not* a number. Thus, for example we have

$$\lim_{x \to 0} \frac{1}{x^2} = \infty,$$

which means that as x approaches zero, the value of $\frac{1}{x^2}$ is increasing beyond all bounds.

The *second* concept of infinity is characterized by the number of elements in a given set. For example, the number of points on the X axis in the interval $0 \le x \le 1$ is denoted by C (for Cantor (1955)). C is the cardinality of the infinite number of points in the given interval. It is called a *transfinite cardinal*. It is a number quite unlike numbers like $1, -27, \pi, 3+4i$. Nevertheless, it is a number and it is intimately associated with continuum modelling, so let us explore its nature by surveying some well known related results (Courant and Robins (1946), Cantor (1955), Zippin (1962)).

To begin, let A and B be two sets. Let an arbitrary element in A be denoted by x and an arbitrary element in B be denoted by y. If to each element x in A one can associate a unique element y in B and to each element y in B one can associate a unique element x in A, then one says that the two sets A and B have been put in one-to-one correspondence. Two sets which have been put into a one-to-one correspondence are said to have the same cardinality, that is, the same number of elements.

Let us examine now the cardinality of the set of points in the interval $0 \leq x \leq 2$. To do this we will proceed as follows. Consider the triangle PQR shown in Fig. 8.1. Let S be the midpoint of PQ and T be the midpoint of PR. Join S and T. Now, let QR have length 2 so that ST has length 1. Indeed, QR is representative of the interval $0 \leq x \leq 2$ while ST is representative of the interval $0 \leq x \leq 1$. Thus, we know that ST has C points. We will now show, strangely enough, that QR also has exactly C points.

Draw any line PL as shown in Fig. 8.1 which intersects QR. Let it intersect ST in x and QR in y. Then to the point x in ST let there correspond the unique point y in QR and to the point y in QR let there correspond the unique point x in ST. Consider *all possible* such lines L. In this fashion ST and QR are put in a one-to-one correspondence. Thus, they have the same cardinality and the number of points in QR is C because the number of points in ST is C. In other words, the number of points in the interval

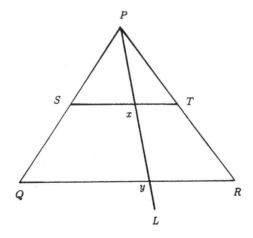

Fig. 8.1.   A one-to-one correspondence.

$0 \leq x \leq 2$ is the same as the number of points in the interval $0 \leq x \leq 1$. However, because the length of the interval $0 \leq x \leq 2$ is twice the length of the interval $0 \leq x \leq 1$, we may surmise, correctly, that the algebra of transfinite cardinals implies that $2C = C$, which is completely different from the algebra satisfied by the real or complex number systems.

Indeed, one can go further and prove that the number of points in the entire real line $-\infty < x < \infty$ is C, that the number of points in the XY plane is C, and that the number of points in XYZ space is also C. Further, it can be shown that if one takes an exceptionally large number of points, like 10^{20} points, on the real line $-\infty < x < \infty$, then these points form a set of measure zero in the line, that is, they are like a zeroth part of the line. In other words, 10^{20} points are of no significance when compared to the C points of the line. Indeed, going from 10^{20} to C is like taking an infinite step, and this is exactly what we are doing when we approximate a molecular configuration by a continuum configuration. Unfortunately, too many of us have been led to believe that 10^{20} is so large that it can be approximated by infinity.

In view of the above revelations about infinity, let us examine a few of the problems that result from continuum modelling.

8.3. The Surface Area Paradox

Consider the section s of the hyperbola $y = 1/x$ on $1 \leq x \leq \infty$, as shown in Fig. 8.2. The section s bounds the crosshatched area in the figure and extends to positive infinity. Rotate s around the X axis to generate the three dimensional surface S shown in Fig. 8.3, which we will call an infinite horn. What we wish to do is calculate both the volume V and surface area A of the horn using the well known formulas from calculus. Hence,

$$V = \pi \int_1^\infty y^2 \, dx$$
$$= \pi \int_1^\infty \frac{1}{x^2} \, dx$$
$$= \pi \left(-\frac{1}{x} \right) \Big|_1^\infty$$
$$= \pi.$$

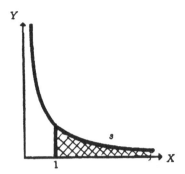

Fig. 8.2.   The curve s.

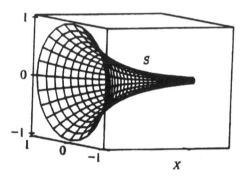

Fig. 8.3.   The surface S.

Thus, the volume is finite. Now,

$$A = 2\pi \int_1^\infty y\sqrt{1 + (y')^2}\, dx$$

$$= 2\pi \int_1^\infty \frac{1}{x}\sqrt{1 + \left(\frac{1}{x}\right)^2}\, dx.$$

But,

$$\sqrt{1 + \left(\frac{1}{x}\right)^2} > 1, \quad x \geq 1,$$

so that

$$A > 2\pi \int_1^\infty \frac{1}{x}\,dx.$$
$$= 2\pi(\log|x|)|_1^\infty$$
$$= \infty.$$

Thus, A is infinite. So, paradoxically, we can fill the inside of S with a finite amount of paint which is not sufficient to paint the outside of S. This result has caused some mathematicians to seek a new definition of surface area.

8.4.   Paradoxes of Zeno

Though the Greek philosopher Zeno formulated several interesting paradoxes relative to the concept of *motion*, let us restrict attention only to two of these. The first paradox is that of the *Moving Arrow*, which proceeds as follows. An archer shoots an arrow into the air from the point A in Fig. 8.4. By direct observation we see the arrow move to the point B in the figure. However, reasons Zeno, consider any fixed time t during which the arrow moved from A to B. At that *instant* in time, the arrow was not moving, since it takes time to move. However, at *every* instant of time during the arrow's flight, the arrow was not moving. Hence the arrow was *never* moving, so that it could not have moved, which contradicts our visual observation.

The second paradox is that of *The Runner and the Tortoise* and proceeds as follows. A runner and a tortoise are to run a race. Because of his lack of speed, the tortoise is given a head start. If one simply observes the race, it follows readily that the runner easily passes the tortoise. However, Zeno now reasons as follows. The runner must first run up to the starting point of the tortoise. This takes time during which the tortoise moves forward to a new position. The runner must then run up to the tortoise's new position, which takes time, during which the tortoise moves forward. The runner must next move up to the tortoise's new position, which takes time, during

Fig. 8.4.   An arrow's parabolic path.

which the tortoise moves forward. Continuing to reason in this fashion, one sees that the runner must always be at a point where the tortoise has been. Thus, since he is always at a point behind the tortoise he can never pass the tortoise.

The paradoxes of Zeno have disturbed mathematicians, physicists and philosophers for 2000 years and have never been resolved satisfactorily. However, most of Zeno's paradoxes disappear immediately if one uses molecular structure instead of a continuum. For example, if the line segment AB were replaced by a linear arrangement of molecules, no paradox can result. The reason is that if one considers any particular *molecule*, then *there is a unique next molecule to its right*. However, if one chooses any *point* in the segment, *there is no unique next point*.

8.5.   A Nonsolvable Problem in Population Genetics

In the study of populations, statisticians and probabilists often employ a distribution function, the most important of which is the standard normal distribution function

$$y = e^{-z^2}, \quad -\infty < z < \infty,$$

whose graph is shown in Fig. 8.5 (Hoel (1971)). The percentage P of the population which lies between any two values $x = c, x = d, c < d$, is given by

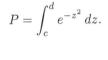

$$P = \int_c^d e^{-z^2}\, dz.$$

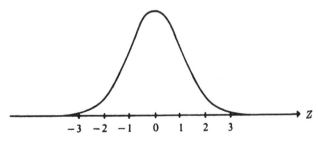

Fig. 8.5.   A standard normal distribution function.

However, no table of integrals gives an antiderivative F for $\int e^{-z^2}\,dz$, so that one cannot evaluate

$$\int_c^d e^{-z^2}\,dz = F(d) - F(c)$$

by the usual rules of calculus. In fact, one can *prove* that no such F exists. Thus the simplest population distribution curve can never yield an exact evaluation. Indeed, it may be pointed out that those functions which do have antiderivatives F which enable one to evaluate their integrals exactly are the exceptions to the rule, since most functions do not. Statisticians and probabilists are forced to use large tables which approximate P for their analyses.

8.6. Time as a Continuum in Fluid Dynamics

Time is not a material object (Whitrow (1961)). It may be that time can be modelled as a continuum, and, indeed, it usually is. This means, for example, that if one models a fluid as a composition of molecules or particles, then their interactions can be modeled as a large system of *ordinary differential equations*, since ordinary differential equations assume the continuity of time. This is the approach we have developed in Chapters 1–7. However, if one models the *fluid* also as a continuum, then the large system of ordinary differential equations can be replaced by a small system of partial differential equations, the most popular of which are the Navier–Stokes equations. However, these equations assume that the flow of the fluid is laminar and noone has as yet shown that these equations can therefore model the most common type of fluid flow, namely, turbulence (see Sections 1.6 and 1.7) Our discrete approach characterizes turbulence as that flow which results when molecular or particle speeds are sufficiently high that the force of repulsion becomes dominant and overwhelms laminar motion.

8.7. Remark

The primary reason for this chapter's discussion is to reinforce the fact that *all* models of natural phenomena are only approximations of the real thing.

References and Additional Sources

R. Agarwal, K. Yun and R. Balakrishnan, "Beyond Navier-Stokes: Burnett equations for flow simulations in continuum-transition regime", AIAA 99-3580, AIAA, 1999.

M. P. Allen, "Algorithms for Brownian dynamics", Mol. Phys., 47, 1982, p. 599.

M. P. Allen and D. J. Tildesley, *Computer Simulation of Liquids*, Clarendon Press, Oxford, 1987.

D. Arnal and R. Michel (Eds.), *Laminar-Turbulent Transition*, Vol. 3, Springer-Verlag, Berlin, 1990.

A. J. Bard, *Integrated Chemical Systems: A Chemical Approach to Nanotechnology*, Wiley, N.Y., 1994.

G. I. Barenblatt, G. Iooss and D. D. Joseph, *Nonlinear Dynamics and Turbulence*, Pitman, Boston, 1983.

G. K. Batchelor, *The Theory of Homogeneous Turbulence*, Cambridge Univ. Press, London, 1960.

P. Berge, Y. Pomeau and C. Vidal, *Order Within Chaos: Towards a Deterministic Approach to Turbulence*, Wiley, N.Y., 1986.

P. S. Bernard, "Transition and turbulence: Basic physics", in *The Handbook of Fluid Dynamics* (Ed. R. W. Johnson), CRC Press, Boca Raton, 1998, pp. 13.1–13.11.

J. P. Boon and S. Yip, *Molecular Hydrodynamics*, Dover, N.Y., 1991.

P. Bradshaw (Ed.), *Turbulence*, Springer, N.Y., 1978.

J. C. Butcher, *Numerical Analysis of Ordinary Differential Equations: Runge-Kutta and Linear Methods*, Wiley, N.Y., 1987.

G. Cantor, *Contributions to the Founding of the Theory of Transfinite Numbers*, Dover, N.Y., 1955.

T. P. Chiang, W. H. Sheu and R. R. Hwang, "Effect of Reynolds number on the eddy structure in a lid-driven cavity", Int. J. Num. Meth. in Fluids, 26, 1998, p. 557.

A. J. Chorin, *Vorticity and Turbulence*, Springer-Verlag, N.Y., 1994.

G. Corliss and Y. F. Chang, "Solving ordinary differential equations using Taylor series", ACM Trans. in Math. Soft., 8, 1982, p. 114.

R. Courant and H. Robbins, *What is Mathematics*, Oxford Univ. Press, N.Y., 1946, pp. 77–88.

B. C. Crandall (Ed.), *Nanotechnology: Molecular Speculations on Global Abundance*, MIT Press, Cambridge, 1996.

C. Crowe, M. Sommerfeld and Y. Tsuji, *Multiphase Flows with Droplets and Particles*, CRC Press, Boca Raton, 1997.

R. J. Donnelly, "Cryogenic Fluid Dynamics", J. Phys.: Condens. Matter, 11, 1999, p. 7783.

K. Eric Drexler, *Engines of Creation*, Anchor, N.Y., 1987.

D. L. Dwoyer, M. Y. Hussaini and R. G. Voigt (Eds.), *Theoretical Approaches to Turbulence*, Springer, N.Y., 1985.

P. W. Egolf, "Difference-quotient turbulence model: A generalization of Prandtl's mixing length theory", Phys. Rev. E, 49, 1994, p. 1260.

R. Eppler and H. Fasel (Eds.), *Laminar-Turbulent Transition*, Vol. I, Springer-Verlag, Berlin, 1980.

D. J. Evans, "Nonlinear viscous flow in a Lennard-Jones fluid", Phys. Lett., 74A, 1979, p. 229.

A. Favre (Ed.), *The Mechanics of Turbulence*, Gordon and Breach, N.Y., 1964.

R. P. Feynman, R. B. Leighton and M. Sands, *The Feynman Lectures on Physics*, Addison-Wesley, Reading, 1963.

R. P. Feynman, "There's plenty of room at the bottom", Invited Lecture, Ann. Meet. APS, Cal. Inst Tech., Pasadena, 1959 (available at www.zyvex.com/nanotech/feynman.html).

C. J. Freitas, R. L. Street, A. N. Findikakis and J. R. Koseff, "Numerical Simulation of Three-Dimensional Flow in a Cavity", Int. J. Num. Meth. in Fluids, 5, 1985, p. 561.

W. Frost and T. H. Moulden (Eds.), *Handbook of Turbulence*, Plenum, N.Y., 1977.

A. Gharakhani and A. Ghoniem, "Three-dimensional vortex simulation of time dependent incompressible internal viscous flows", J. Comp. Phys., 134, 1997, p. 75.

C. G. Gray and K. E. Gubbins, *The Theory of Molecular Fluids. I. Fundamentals*, Clarendon Press, Oxford, 1984.

D. Greenspan, "A molecular mechanics-type approach to turbulence", Mathl. Comput. Modelling, 26, 1997(a), p. 85.

D. Greenspan, *Particle Modeling*, Birkhauser, Boston, 1997(b).

D. Greenspan, "Molecular cavity flow", Fluid Dynamics Research, 25, 1999(a), p. 37.

D. Greenspan, "A molecular mechanics study of the development of turbulence in a dense gas", Physica Scripta, 60, 1999(b), p. 242.

J. O. Hinze, *Turbulence*, McGraw-Hill, N.Y., 1975.

J. O. Hirschfelder, C. F. Curtiss and R. B. Bird, *Molecular Theory of Gases and Liquids*, Wiley, N.Y., 1967.

H. C. Hoch, H. G. Craighead and L. W. Jelenski (Eds.), *Nanofabrication and Biosystems: Integrating Materials, Science, Engineering, and Biology*, Camb. Univ. Press, N.Y., 1996.

R. W. Hockney and J. W. Eastwood, *Computer Simulation Using Particles*, McGraw-Hill, N.Y., 1981.

P. G. Hoel, *Elementary Statistics* (3rd Ed.), Wiley, N.Y., 1971, p. 103.

R. A. Horne, Encyclopedia Brittanica, Micropaedia, Vol. 19, p. 634, 15th Edition, 1981.

E. Hopf, "A mathematical example displaying features of turbulence", Comm. Pure Appl. Math., 1, 1948, p. 303.

R. Kobayashi (Ed.), *Laminar-Turbulent Transition*, Vol. 4, Springer-Verlag, Berlin, 1995.

A. N. Kolmogorov, "Toward a more precise notion of the structure of the local turbulence in a viscous fluid at elevated Reynolds numbers", (see A. Favre, above), p. 447.

J. Koplik and J. R. Banavar, "Physics of fluids at low Reynolds number — A molecular approach", Computers in Physics, 12, 1998, p. 424.

A. Korzeniowski and D. Greenspan, "Microscopic turbulence in water", Mathl. Comput. Modelling, 23, 1996, p. 89.

J. R. Koseff and R. L. Street, "Visualization studies of a shear driven three-dimensional recirculating flow", J. Fluids Eng., 106, 1984, p. 21.

V. V. Kozlov (Ed.), *Laminar-Turbulent Transition*, Vol. 2, Springer-Verlag, Berlin, 1985.

O. A. Ladyzhenskaya, *The Mathematical Theory of Viscous, Incompressible Flow*, (2nd Ed.), Gordon and Breach, N.Y., 1969, pp 5–6.

L. D. Landau, "On the problem of turbulence", Dokl. Akad. Nauk USSR, 44, 1944, p. 311.

B. E. Launder and D. P. Tselepidakis, "Progress and paradoxes in modeling near-wall turbulence", Turbulent Shear Flows, 8, 1993, pp. 81–96.

B. E. Launder and D. B. Spalding, *Mathematical Models of Turbulence*, Academic, N.Y., 1972.

M. Lesieur, *Turbulence in Fluids*, (3rd Ed.), Kluwer Acad., Dordrecht, 1997.

P. A. Libby, *Introduction to Turbulence*, Taylor and Francis, Washington, D.C., 1996.

J. Lumley, "Some comments on turbulence", Phys. Fluids A. 4, 1992, pp. 203–211.

W. V. R. Malkus, "Summer program notes on geophysics and fluid dynamics", Woods Hole Ocean. Inst., Woods Hole, MA, 1960.

M. H. March and M. P. Tosi, *Atomic Dynamics in Liquids*, Dover, N.Y., 1991.

N. C. Markatos, "The mathematical modelling of turbulence", Appl. Math. Model., 10, 1986, p. 190.

W. L. Masterton and E. J. Slowinski, *Chemical Principles* (2nd Ed.), Saunders, Phila., 1969, p. 96.

W. D. McComb, *The Physics of Turbulence*, Oxford, N.Y., 1990.

H. D. Megaw, *Crystal Structure: A Working Approach*, Saunders, Phila, 1973.

F. Pan and A. Acrivos, "Steady flows in a rectangular cavity", J. Fluid Mech., 28, 1967, p. 643.

C.-Y. Perng and R. L. Street, "Three-dimensional unsteady flow simulations: Alternative strategies for a volume-averaged calculation", Int. J. Num. Meth. in Fluids, 9, 1989, p. 341.

L. Prandtl, "Über die ausgebildete turbulenz", *Zamm*, 5, 1925, p. 136.

J. G. Prowles, "The liquid-vapor coexistence line for Lennard–Jones type fluids", Physica, A126, 1984, p. 289.

A. H. Rahman and F. H. Stillinger, "Molecular dynamics study of liquid water", J. Chem. Phys., 55, 1971, p. 3336.

D. C. Rapaport, *The Art of Molecular Dynamics Simulation*, Cambridge Univ. Press, Cambridge, 1995, Ch. 8.

E. Regis, *Nano: The Energy Science of Nanotechnology*, Little and Brown, N.Y., 1996.

P. J. Roache, *Computational Fluid Dynamics*, Hermosa, Albuquerque, 1972, p. 204.

D. Ruelle and F. Takens, "On the nature of turbulence", Comm. Math. Phys., 20, 1971, p. 167.

H. Schlichting, *Boundary Layer Theory*, McGraw-Hill, N.Y., 1960.

C. G. Speziale and R. M. C. So, "Turbulence modeling and simulation" in *The Handbook of Fluid Dynamics* (Ed. R. W. Johnson), CRC Press, Boca Raton, 1998, pp. 14.1–14.111.

M. M. Stanisi'c, *The Mathematical Theory of Turbulence*, Springer, N.Y., 1988.

F. H. Stillinger and A. Rahman, "Improved simulation of liquid water by molecular dynamics", J. Chem. Phys., 60, 1974, p. 1545.

V. L. Streeter, *Fluid Mechanics* (3rd Ed.), McGraw-Hill, N.Y., 1962, p. 533.

I. M. Svishchev and P. G. Kusalik, "Structure in liquid water: A study of spatial distribution functions", J. Chem. Phys., 99, 1993, pp. 3049–3058.

G. I. Taylor, "Diffusion by continuous movements", Proc. London Math. Soc. A, 20, 1921, p. 196.

M. Van Dyke, *An Album of Fluid Motion*, Parabolic Press, Stanford, CA, 1982.

T. von Karman, "The fundamentals of the statistical theory of turbulence", J. Aero. Sci., 4, 1937b, p. 131.

J. D. Weeks, D. Chandler and H. C. Andersen, "Role of repulsive forces in determining the equilibrium structure of simple liquids", J. Chem. Phys., 54, 1971, p. 5237.

G. J. Whitrow, *The Natural Philosophy of Time*, Harper and Row, N.Y., 1961.

V. Yakhot and S. Orszag, "Renormalization group analysis of turbulence I: Basic theory", J. Sci. Comp., 1, 1986, pp 3–51.

N. J. Zabusky (Ed.), *Topics in Nonlinear Physics*, Springer, N.Y., 1968.

Y. Zang, R. L. Street and J. Koseff, "Large eddy simulation of turbulent cavity flow using a dynamic subgrid-scale model", FED-Vol. 162, Eng. Appl. Large Eddy Simul., ASME, 1993.

L. Zippin, *Uses of Infinity*, Random House, N.Y., 1962.

Index